110 IC Timer Projects

110 IC Timer Projects

Jules H. Gilder

HAYDEN BOOK COMPANY, INC.
Rochelle Park, New Jersey

To my wife Miriam for her encouragement and support and to my children Gayle and Jordan who sacrificed many happy hours with their father so that this book could be written.

Library of Congress Cataloging in Publication Data

Gilder, Jules H   1947–
    110 IC timer projects.

    1. Timing circuits. I. Title.
TK7868.T5G54   621.381'73'7   78-23360
ISBN 0-8104-5688-5

6   7   8   9   PRINTING

82 83 84 85 86 87   YEAR

preface

110 IC Timer Projects is a sourcebook of applications for the ubiquitous 555 IC timer. The basic operation of the device is detailed to provide both engineer and hobbyist with design ideas for interesting and useful circuits that have found application in everyday life.

The book is divided into eight chapters. The first introduces the 555 timer, describes how the basic integrated circuit works, and presents IC pin comparisons of the different packages. It also includes the basic data sheet for the device.

The next three chapters cover the basic modes of operation of the 555. Chapter Two describes how it is used as a monostable multivibrator, and several variations of the basic monostable circuit are presented. Chapter Three details the astable mode of operation. Chapter Four explains how the 555 can be used in applications for which it was not originally intended—how this basically analog device can be used as a variety of different logic devices. After reading these three chapters, you should come away with a fundamental understanding of what the 555 is and how to use it.

The remaining four chapters of the book cover applications in instruments, in the automobile, in alarm and control circuitry, and in power supplies and converters. The wide variety of applications of this tiny electronic component point up how electronics is becoming an increasing part of our everyday lives.

I would like to thank all the publishers who gave permission to include material that originally appeared elsewhere. Footnotes to selected articles cite these original sources.

Jules H. Gilder

contents

getting to know the timer

Integrated circuit (IC) timers are probably the most versatile ICs available today. They can provide precise timing intervals ranging from microseconds to hours. They can also be used as easily controllable, inexpensive oscillators. Applications range from interface circuitry to control systems and alarms.

The most popular of the IC timers available is the 555 timer. It generally comes in an 8-pin minidip package (Fig. 1-1A) but is also available in an 8-pin round TO-99 can (Fig. 1-1B). In addition, a 14-pin dual in-line package (Fig. 1-1C) containing two 555 IC timer chips is available. This is generally designated as the 556 timer, although some manufacturers do use other number designations (XR-2556 and D555) with different pinout configurations (Fig. 1-1D).

The 555 timer is basically a very stable IC that is capable of being operated either as an accurate bistable, monostable, or astable multivibrator. These three basic modes of operation make the 555 one of the most useful ICs since the operational amplifier.

The 555 IC timer is composed of 25 transistors, 2 diodes, and 16 resistors (Fig. 1-2) and might appear to be quite complex, but, as can be seen from the block diagram in Fig. 1-3, the device is functionally quite simple. The circuitry in the IC timer is arranged to form two comparators, a flip-flop, two control transistors, and a high-current output stage.

The comparators are actually operational amplifiers that compare input voltages to internal reference voltages that are generated by a voltage divider consisting of three 5,000-ohm resistors. The references provided by this divider are two-thirds of the supply voltage (V_{cc}) and one-third of V_{cc}. When the input voltage to either one of the comparators is higher than the reference voltage for that comparator, the operational amplifier goes into saturation and thus produces a signal that is used to trigger the flip-flop. And the flip-flop controls the output state of the timer.

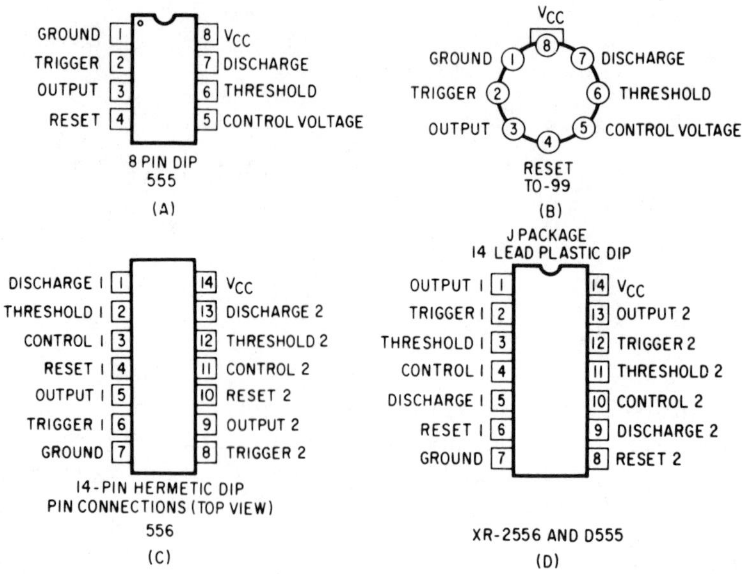

Fig. 1-1.   Package configurations.

A quick look at what each pin on the 555 IC does will help provide a clearer understanding of the 110 circuits that follow.

Pin 1. This is the *ground* pin and gets connected to the negative side of the voltage supply.

Pin 2. This is the *trigger* input. When a negative-going pulse causes the voltage at this point to drop below one-third of the V_{cc}, the comparator to which this input is connected causes the flip-flop to change state, causing the output level to switch from low to high. The trigger pulse must be of shorter duration than the time interval determined by the external R and C. If this pin is held low longer than that, the output will remain high until the trigger input is driven high again.

Pin 3. This is the *output* pin. It is capable of sinking or sourcing a load that requires up to 200 mA of current and can drive TTL circuits. The output voltage available at this pin is approximately equal to the V_{cc} applied to pin 8 minus 1.7 V.

Pin 4. This is the *reset* pin. It is used to reset the flip-flop that controls the state of output pin 3. The pin is activated when a voltage level anywhere between 0 and 0.4 V is applied to the pin. The reset pin will force the output to go low no matter what state the other inputs to the flip-flop are in. To prevent unwanted resetting of the output, pin 4 should be connected along with pin 8 to the positive side of V_{cc} when not in use.

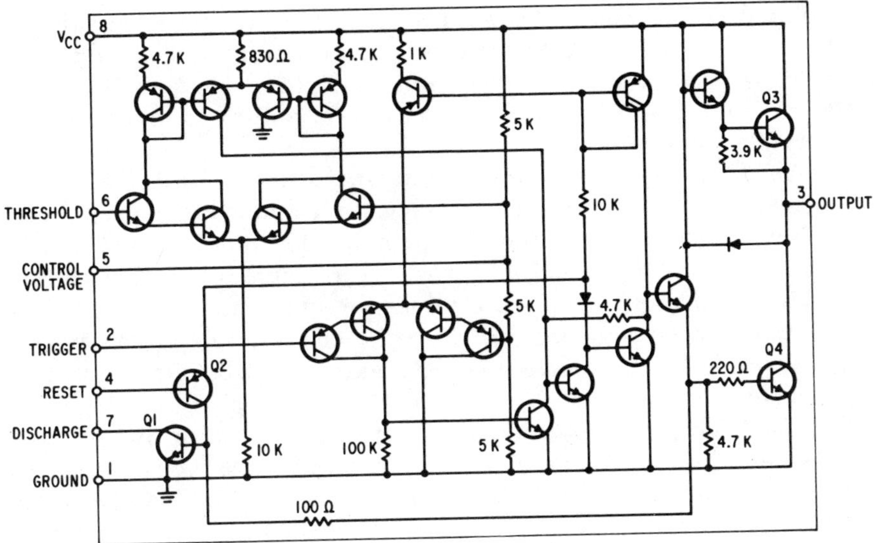

Fig. 1-2. 555 timer schematic.

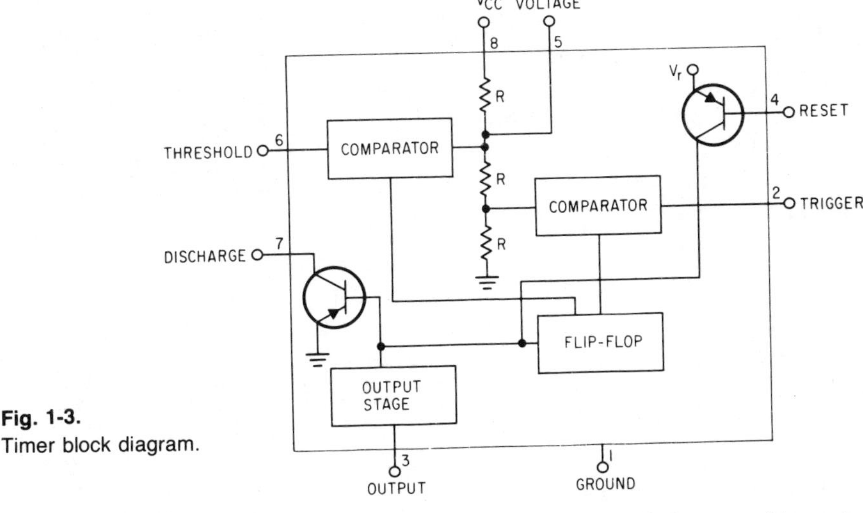

Fig. 1-3.
Timer block diagram.

Pin 5. This is the *control voltage input.* By applying a voltage to this pin, it is possible to vary the timing of the device independently of the RC network. The control voltage tmay be varied from 45 to 90% of the V_{cc} in the monostable mode, making it possible to control the width of the output pulse independently of RC. When it is used in the astable mode, the control voltage can be varied from 1.7 V to the full V_{cc}. Varying the voltage in the astable mode will produce a frequency-modulated (FM) output. This pin is connected to the internal voltage divider so a voltage measurement between it and ground should read two-thirds of the voltage applied to pin 8. If, as in most applications, this pin is not used, it should be bypassed to ground to maintain immunity from noise.

Pin 6. This is the *threshold* input. It resets the flip-flop and consequently drives the output low if the voltage applied to it rises above two-thirds of the value of the voltage applied to pin 8. In addition to the voltage level, a current of at least 0.1 μA must be supplied to this pin. This threshold current determines the maximum value of resistance that can be connected between the positive side of the supply and this pin. For 15 V operation, the maximum value of resistance is 20 megohms (MΩ).

Pin 7. This is the *discharge* pin. It is connected to the collector of an npn transistor. The emitter of the transistor is connected to ground, so that when the transistor is turned "on," pin 7 is effectively shorted to ground. Usually the timing capacitor is connected between pin 7 and ground and is discharged when the transistor is turned "on."

Pin 8. This is the *power supply* pin and is connected to the positive side of the supply. The voltage applied to this pin may vary from 4.5 to 16 V for commercial devices. Selected devices that operate at voltages as high as 18 V are available.

The pin numbers and functions for the 8-pin minidip package and the round TO-99 can are identical. The equivalent pin numbers for the 14-pin DIP dual timer are shown in Table 1-1. Electrical characteristics for the 555 timer are shown in Table 1-2, while typical operating curves are shown in Fig. 1-4.

Table 1-1. Pinout Comparison Chart

| | 555 Timer | 556 | | D555 and XR-2556 | |
| | | Timer No. 1 | Timer No. 2 | Timer No. 1 | Timer No. 2 |
Function	*Pin No.*	*Pin No.*	*Pin No.*	*Pin No.*	*Pin No.*
Ground	1	7		7	
Trigger	2	6	8	2	12
Output	3	5	9	1	13
Reset	4	4	10	6	8
Control voltage	5	3	11	4	10
Threshold	6	2	12	3	11
Discharge	7	1	13	5	9
V_{cc}	8	14		14	

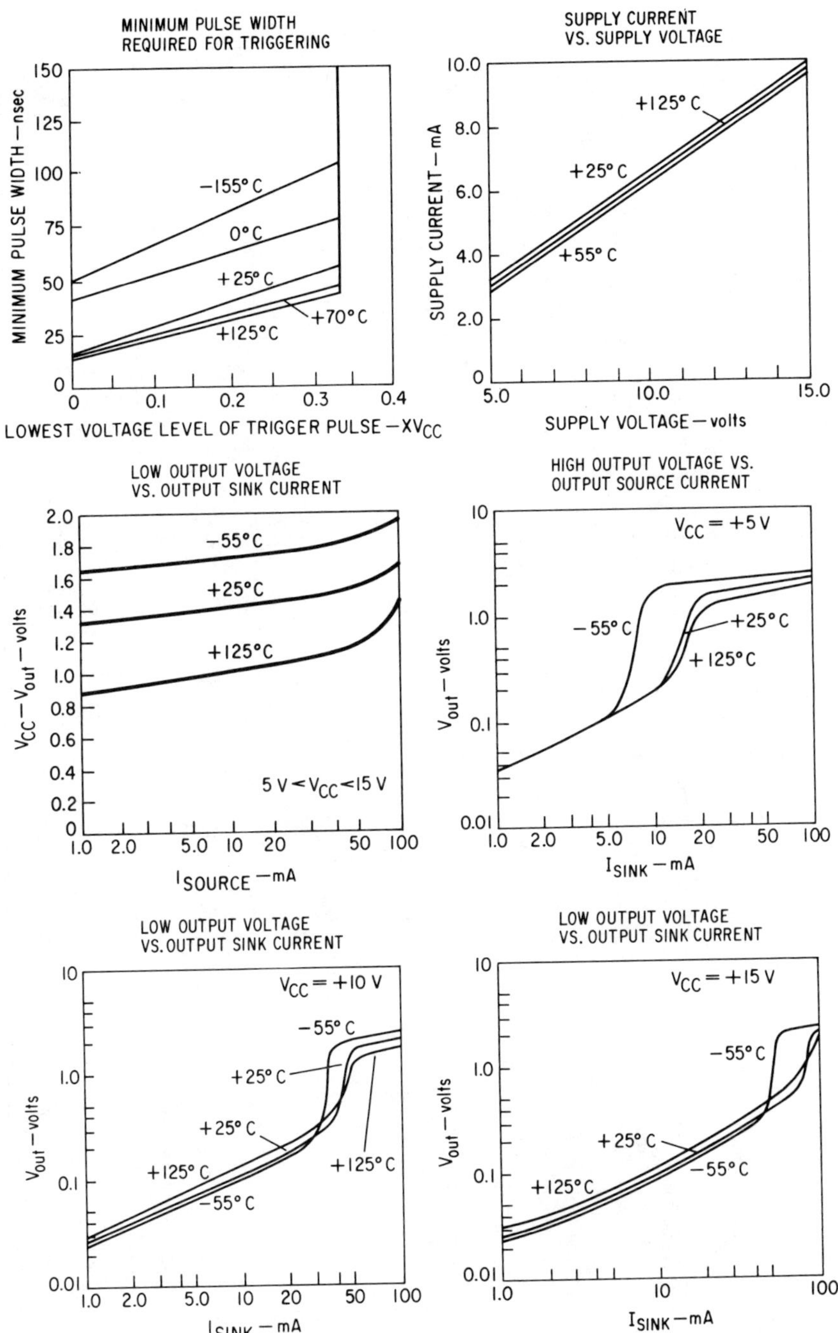

Fig. 1-4. Typical operating curves.

Table 1-2. Electrical Characteristics: $T_A = 25°C$, $V_{cc} = +5$ to $+15$ V unless otherwise specified

Parameter	Test conditions	SE 555			NE 555			Units
		Min	Typ	Max	Min	Typ	Max	
Supply voltage		4.5		18	4.5		16	V
Supply current	$V_{cc} = 5V\ R_L = \infty$		3	5		3	6	mA
	$V_{cc} = 15V\ R_L = \infty$		10	12		10	15	mA
	Low state; Note 1							
Timing error	$R_A, R_B = 1k\Omega$ to $100\ k\Omega$							
Initial accuracy	$C = 0.1\ \mu F$; Note 2		0.5	2		1		%
Drift with temperature			30	100		50		ppm/°C
Drift with supply voltage			0.05	0.2		0.1		%/Volt
Threshold voltage			⅔			⅔		$\times V_{cc}$
Trigger voltage	$V_{cc} = 15$ V	4.8	5	5.2		5		V
	$V_{cc} = 5$ V	1.45	1.67	1.9		1.67		V
Trigger current			0.5			0.5		μA
Reset voltage		0.4	0.7	1.0	0.4	0.7	1.0	V
Reset current			0.1			0.1		mA
Threshold current	Note 3		0.1	.25		0.1	.25	μA
Control voltage level	$V_{cc} = 15$ V	9.6	10	10.4	9.0	10	11	V
	$V_{cc} = 5$ V	2.9	3.33	3.8	2.6	3.33	4	V
Output voltage drop (low)	$V_{cc} = 15$ V							
	$I_{sink} = 10$ mA		0.1	0.15		0.1	.25	V
	$I_{sink} = 50$ mA		0.4	0.5		0.4	.75	V
	$I_{sink} = 100$ mA		2.0	2.2		2.0	2.5	V
	$I_{sink} = 200$ mA		2.5			2.5		V
	$V_{cc} = 5$ V							
	$I_{sink} = 8$ mA		0.1	0.25				V
	$I_{sink} = 5$ mA					.25	.35	V
Output voltage drop (high)	$I_{source} = 200$ mA		12.5			12.5		V
	$V_{cc} = 15$ V							
	$I_{source} = 100$ mA							
	$V_{cc} = 15$ V	13.0	13.3		12.75	13.3		V
	$V_{cc} = 5$ V	3.0	3.3		2.75	3.3		V
Rise time of output			100			100		nsec
Fall time of output			100			100		nsec

Notes: [1]Supply current when output high typically 1 mA less. [2]Tested at $V_{cc} = 5$ V and $V_{cc} = 15$ V. [3]This will determine the maximum value of $R_A + R_B$. For 15-V operation, the max total R = 20 MΩ.

chapter
two

monostable
circuits

2.1   the basic monostable

The monostable, or one-shot, multivibrator is a circuit that produces a single pulse when triggered. The circuit of the 555 hooked up to operate as a monostable is shown in Fig. 2-1. The width of the pulse produced, sometimes called time delay, is determined by two external components, R_A and C. The external capacitor is initially held in the discharge state by a transistor inside the timer, which shorts it to ground. The timing cycle begins when a negative trigger pulse is applied to pin 2. This pulse sets the internal flip-flop (changes its state), removing the base bias to the discharge transistor and thus removing the short circuit across the external capacitor. The voltage across the capacitor then starts to increase exponentially with a time constant (time required to charge to 63% of the final voltage, or discharge to 37% of the initial voltage) of $t = R_A C$. When the voltage across the capacitor reaches a threshold value—in this case $\frac{2}{3} V_{cc}$ as determined by the internal voltage divider—a comparator resets the flip-flop (restores it to its original state). This in turn rapidly discharges the external capacitor and drives the output to a low state.

The circuit triggers when the negative-going pulse applied to pin 2 reaches $\frac{1}{3} V_{cc}$. (The trigger pulse must be narrower than the desired output pulse.) Once triggered, the circuit will remain in this state until

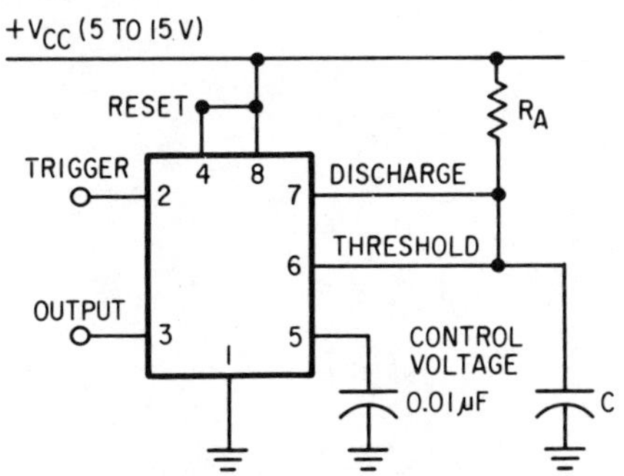

Fig. 2-1.   Monostable multivibrator.

the set time has elapsed, even if it is triggered again during this interval. But it can be reset if a negative pulse is applied to reset pin 4. The charging rate and the threshold level of the comparator are both directly proportional to supply voltage V_{cc}, and thus the timing interval is independent of the supply voltage.

The amount of time that the output of the monostable remains in the high state is given by $t = 1.1R_{A}C$, where R_{A} is in megohms, C is in microfarads, and t is in seconds. The graph shown in Fig. 2-2 can be used to quickly determine component values.

2.2   resettable monostable

By making a very simple change to the circuit shown in Fig. 2-1, it is possible to build a resettable monostable multivibrator that can be stopped in the middle of a cycle and restarted again (Fig. 2-3). This is done by connecting the trigger input to the reset (pins 2 and 4). Now, if a negative-going pulse is applied to these two inputs, the

Fig. 2-2.   Monostable time-delay graph.

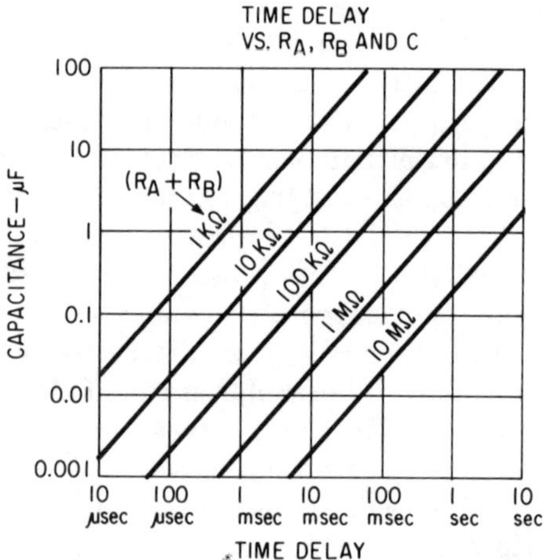

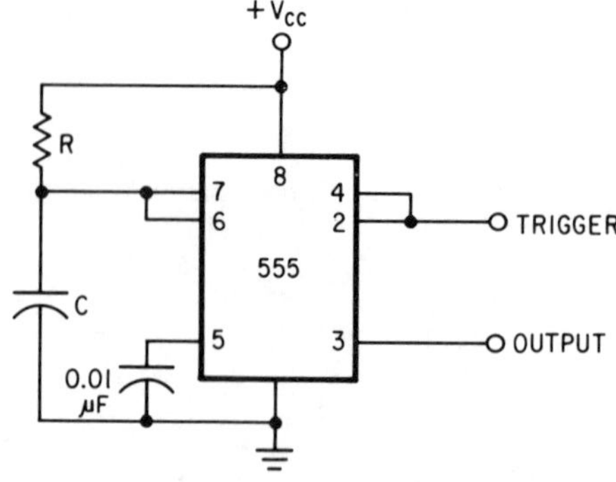

Fig. 2-3.   Resettable monostable.

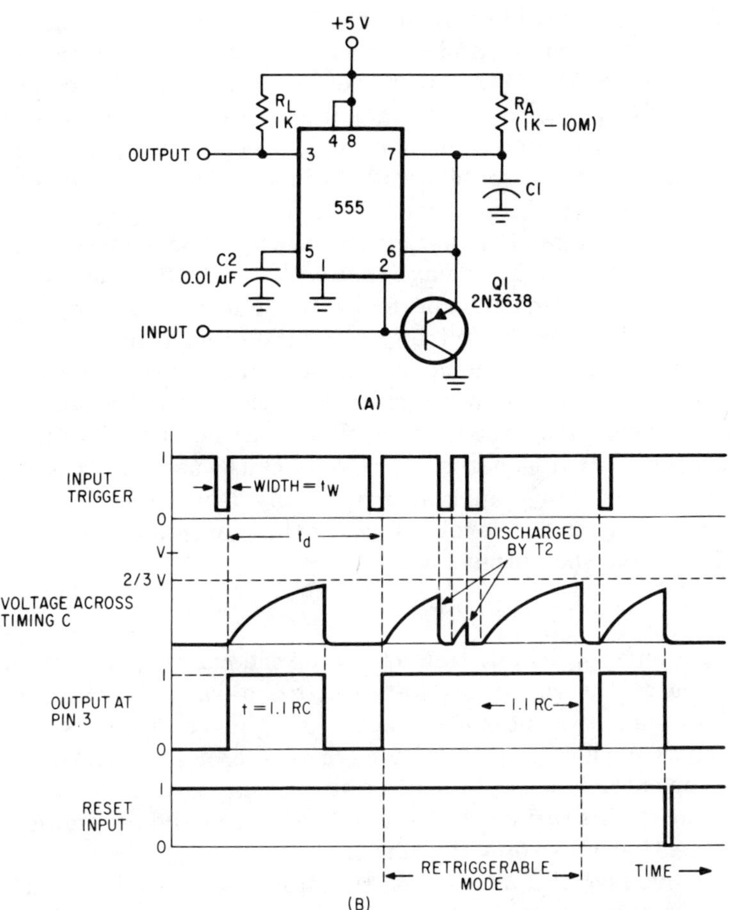

Fig. 2-4.   (A) Retriggerable monostable. (B) Timing waveforms of retriggerable monostable.

timing capacitor will be discharged, the output will go low, and the cycle will start all over again, beginning on the positive edge of the reset pulse. As with the conventional monostable, the output pulse width of this circuit is determined by the formula $t = 1.1RC$.

2.3   retriggerable monostable*

It is not always necessary to drive the output of the monostable low when it is retriggered. By adding a pnp transistor to the basic monostable multivibrator, it is possible to build a monostable that can be retriggered without resetting the output (Fig. 2-4A).

* Mattis, J., "Missing Pulse Detector Reacts to 100 ns Pulse Widths," *Electronic Design*, Nov. 23, 1972, p. 174.

Basically, what happens in this case is that the timing capacitor connected between pin 6 and ground is discharged by the trigger pulse, which arrives after the initial trigger but during the monostable time delay—1.1RC. If the retriggering pulse arrives during this period, the pnp transistor turns on and shorts out the capacitor. After the pulse disappears, the transistor shuts off again and then the capacitor starts to charge again through its series resistor.

The advantage of using a retriggerable monostable is that it is possible to obtain longer delays than with a regular monostable. The total delay is the sum of the normal 1.1RC plus the trigger pulse width (t_w) and the delay between triggers (t_d) (See Fig. 2-4B).

In practical applications, the width of the monostable output pulse (1.1RC) is chosen so that it is about 30% greater than the time between trigger pulses. This ensures that there will be enough time for the timing capacitor to discharge. This circuit can be used as a missing pulse detector or as a "low" rate alarm, since any decrease in the triggering input below the design level will allow the monostable to complete its cycle and drive the output low.

2.4   long-delay monostable*

The pulse width, or time delay, of the conventional monostable is limited by the values of R and C. It is always desirable to use as small a capacitor as possible to achieve highest accuracy. Large-value electrolytic capacitors generally have very broad tolerances. But if a small capacitor is used, a large resistor must be used to compensate. The value of this resistor is limited to 20 MΩ by the minimum current of 0.1 μA that must flow through it.

To increase the amount of resistance permitted between pin 6 and V_{cc}, a programmable unijunction transistor, two resistors and a capacitor are added to the basic monostable circuit (Fig. 2-5). The programmable unijunction transistor oscillates at about 1 Hz. This superimposes 0.1-V negative spikes on the dc level of pin 5, the control voltage input. As the voltage across the timing capacitor reaches the trip point, the threshold appears 0.1 V lower each time a pulse arrives at pin 5. The capacitor then charges to 0.1 V above the threshold, and it, rather than the timing resistor, supplies the necessary current to switch the IC.

The negative spikes with their short duty cycle have little effect on the charging current. This technique makes it possible to use resistors as large as 200 MΩ to charge the capacitor, a 10-fold increase over what is normally allowed. A transistor connected in parallel with the timing capacitor can be used to reset the circuit.

---

* Roe, B. C., "Unijunction Oscillator Helps Increase Range of Monolithic Timer Without Use of Big Capacitors," *Electronic Design*, Oct. 25, 1973, p. 114.

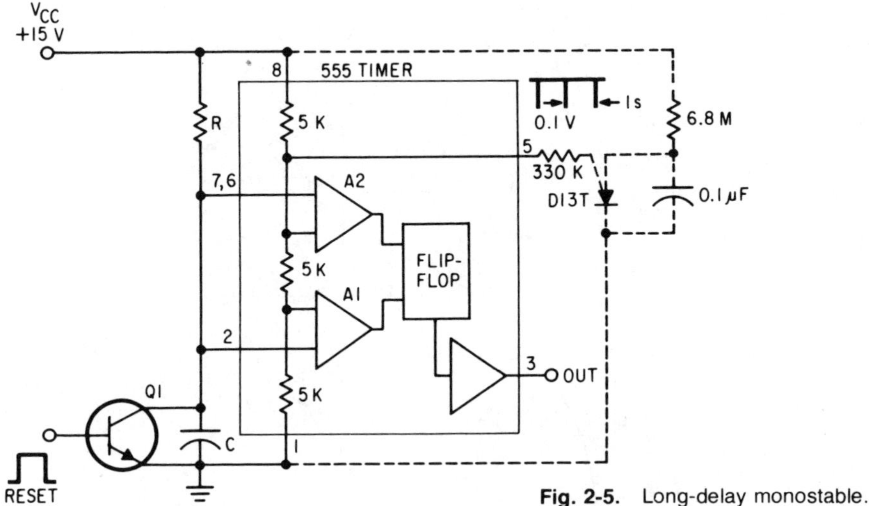

Fig. 2-5.   Long-delay monostable.

2.5   longer-delay monostable*

Whereas the previous circuit increased the maximum attainable delay for the monostable 10 times, this circuit (Fig. 2-6) increases the maximum delay 100 times. In this circuit, the R and C delay elements form part of an integrator, with a CA3140 operational amplifier as the active element.

After the 555 is triggered, capacitor C charges at a rate equal to V_r/RC. With V_r set at 0.1 V, a time equal to 100 RC is needed for C to charge to 10 V, which is $\frac{2}{3}V_{cc}$. At this voltage, the CA3140 output is strobed low and C discharges through diode D.

For high values of R, diode D must have low leakage such as provided by the gate channel diode of a field-effect transistor (FET)—such as the Siliconix E202. The maximum leakage current of the E202 FET is only 100 pA (1 pA = 1×10^{-12} A) at 20 V. The maximum forward current is 50 mA.

Since reference voltage V_r varies with the supply voltage, timing is nearly independent of the supply voltage. An important thing to note is that if a supply voltage of 5 V is being used, the output of the CA3140 will not be able to reach the level of $\frac{2}{3}V_{cc}$. It will therefore be necessary to lower the triggering point of the 555 by adding external resistors.

The charging-time curves of capacitor C with and without the integrator are shown in Fig. 2-6B. The reference voltage used in this integrator was 0.5 V.

* Olthoff, W., "Low-Cost Integrator Multiplies 555 Timer's Delay Range 100 Times," *Electronic Design*, April 26, 1977, p. 120.

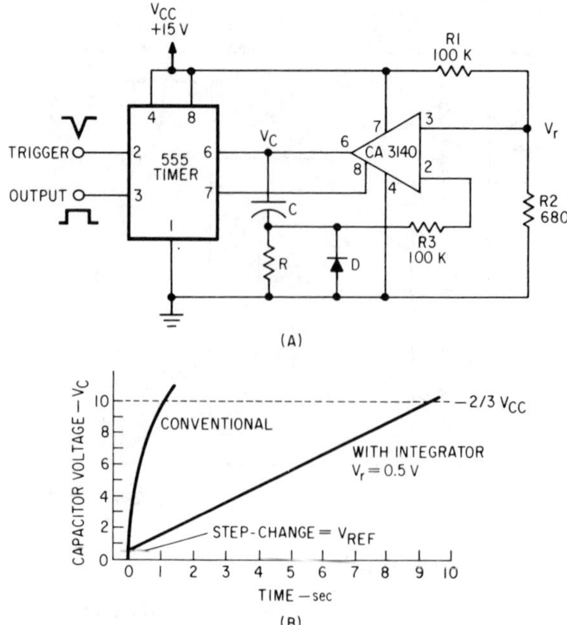

Fig. 2-6. Longer-delay monostable.

2.6   positively triggered monostable*

The standard monostable configuration has negative-level triggering and positive output pulses. But the same device may also be used with positive-going triggering to provide negative output pulses.

The new configuration (Fig. 2-7) may be said to be a mirror image of the standard one-shot, and its timing equations are identical to the original because of the symmetry of the internal comparator levels. Thus, the pulse width remains independent of the power supply and is equal to 1.1RC.

The positive-trigger monostable not only uses a minimum of components, but also offers the significant bonus of two independent outputs, one of which is TTL compatible.

At start-up, the trigger input to pin 6 is low, C is discharged, and pin 2 is momentarily held low. This condition trips the internal latch and forces output 1 (pin 3) high. This causes the timing capacitor to start charging up. The capacitor charges up to a value near V_{cc} and the circuit remains in this state.

With a positive-going input trigger greater than ⅔V_{cc}, which is the threshold level for pin 6, output 1 is forced low. This action effec-

* Jung, W. G., "555 One-Shot Circuit Features Negative Output with Positive Triggering," *Electronic Design*, Aug. 16, 1976, p. 98.

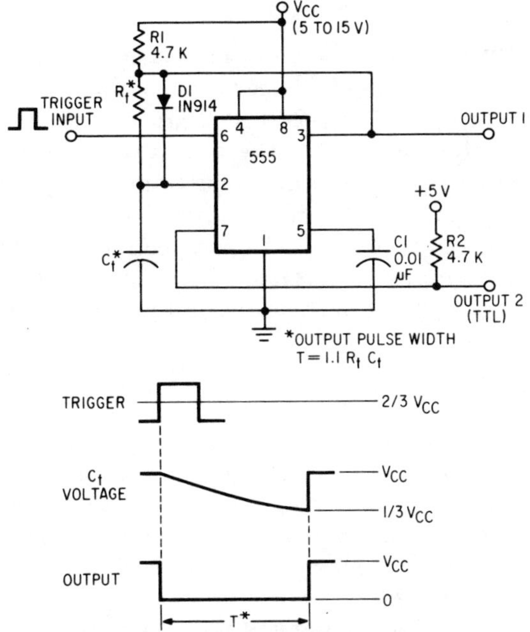

Fig. 2-7.   Positively triggered monostable.

tively grounds the point between R_t and R1. The timing capacitor then begins to discharge through R_t until the voltage level reaches ⅓ V_{cc}, which is the pin 2 threshold level. When this level is reached, output 1 switches high again, quickly recharging the capacitor to V_{cc} via the diode and R1.

An advantage of this new configuration is the availability of pin 7, which can be used as an open-collector output (output 2). Pin 7 may be used as a 5-V TTL drive, regardless of the timer's supply voltage. It has a drive-current capability that is similar to that of the regular output (pin 3).

Both outputs may be used simultaneously, but if the best possible timing accuracy is desired, only output 2 should be used. This is because loading of pin 3 can affect timing, particularly at low supply voltages. If desired, R1 may be deleted with a slight sacrifice in timing accuracy. The reason for this decrease in performance is that in the absence of R1, pin 3 cannot pull up to V_{cc}, and thus capacitor C cannot charge fully between output pulses.

Like the standard monostable, this mirror-image configuration requires an input trigger pulse whose width is less than that of the output pulse. Unlike the standard monostable, the range of values for R is somewhat limited because the bias current at pin 2 is five times larger than that of pin 6.

2.7   variable-pulse-width monostable

By substituting a potentiometer for a capacitor in the basic mono-stable circuit, it is possible to produce a monostable circuit whose output pulse width can be varied (Fig. 2-8). The capacitor to be changed is not the timing capacitor, but rather the capacitor from pin 5 to ground, which is generally used to provide noise immunity protection.

By connecting the center wiper arm of the potentiometer to pin 5 and the two ends to ground and V_{cc}, respectively, it is possible to use the control voltage input of the timer to adjust the width of the output pulse. The voltage applied to this input may be varied from 45 to 90% of V_{cc}.

2.8   linear-sweep monostable

In addition to getting a rectangular pulse from a monostable, it is also possible to get a sweep voltage, or voltage ramp, by using the voltage that appears across the timing capacitor (junction of pins 6 and 7). This sweep voltage is exponential, but by adding a transistor and two resistors to the basic monostable configuration, the ramp can be linearized (see Fig. 2.9A).

To get a linearly increasing voltage, it is only necessary to charge the timing capacitor through a constant-current source instead of a resistor. The charging time for such a circuit is given by $t = 0.67V_{cc}(C)/I$. The charging current (I) in turn is determined by the relationship: $I = (V_{cc} - V_E)/R_E$. If V_E is replaced by its equivalent, which is $V_B - V_{BE}$, then the equation becomes: $I = (V_{cc} - V_B - V_{BE})/R_E$.

Now, if the voltage divider determined by R1 and R2 is adjusted so that V_{BE} is negligible compared with $V_{cc} - V_B$, then the equation for the charging current becomes: $I = (V_{cc} - V_B)/R_E$. However, since V_B is a fixed fraction of the supply voltage and thus directly proportional to it, the equation for the charging current can be written as: $I = (V_{cc} - kV_{cc})/R_E$, where constant k is equal to V_B/V_{cc}.

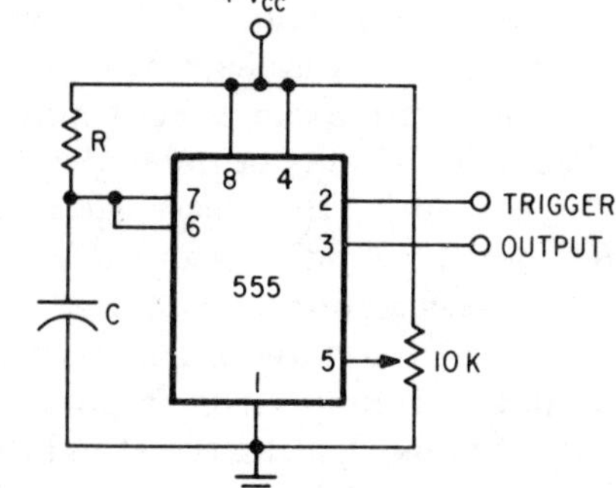

Fig. 2-8.   Variable-pulse-width monostable.

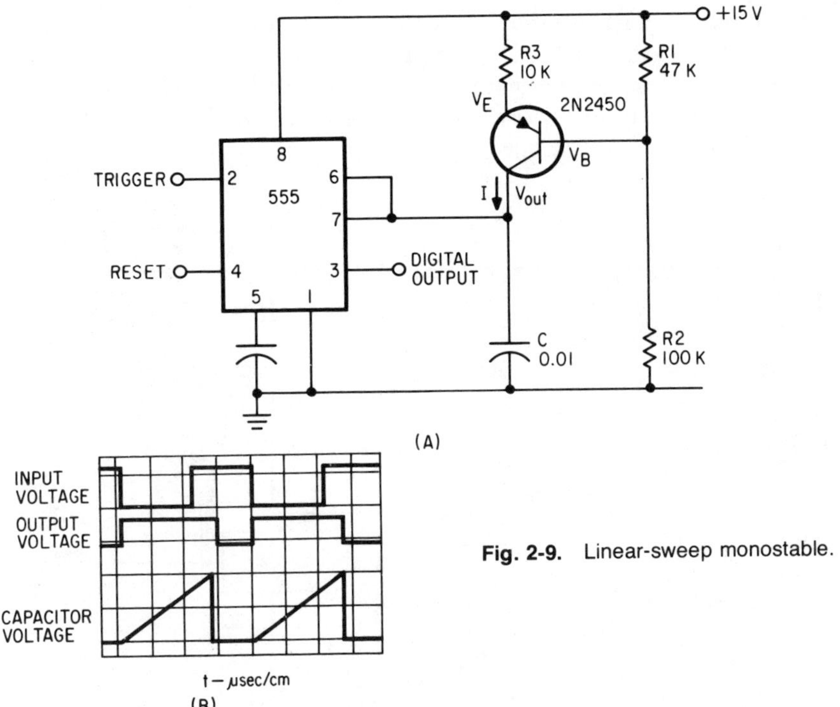

Fig. 2-9. Linear-sweep monostable.

If this equation for the charging current is now substituted into the original equation for the charging time, the result is: $t = 0.67(R_E)(C)/(1-k)$.

Because of the 555's internal voltage divider and comparators, it is very important that the capacitor voltage be able to reach $\frac{2}{3} V_{cc}$. To insure this, a well-regulated supply or a separate, higher-voltage source must be used for the current source. Another important point to note is that the constant current provided by the transistor must be larger than 1 μA.

2.9   compensated monostable*

Often it is desirable to have a monostable multivibrator whose output pulse width can be manually selected via a multiposition rotary switch. What is most often done is to simply switch in different values of timing resistance while leaving the timing capacitor the same. However, if wide pulse widths are to be produced, the accuracy obtainable is severely limited, because close tolerance, high-value timing capacitors are expensive and difficult to obtain.

* Martens, A. E., "Switch Selects Accurate Time Delays," *EDN*, Nov. 5, 1974, p. 62.

If only a small number of switchable outputs are required, this problem can be easily solved by simply using trimmer resistors in series with the regular timing resistor to adjust the individual pulse widths. If, on the other hand, many options are desired, this can become a cumbersome and costly approach.

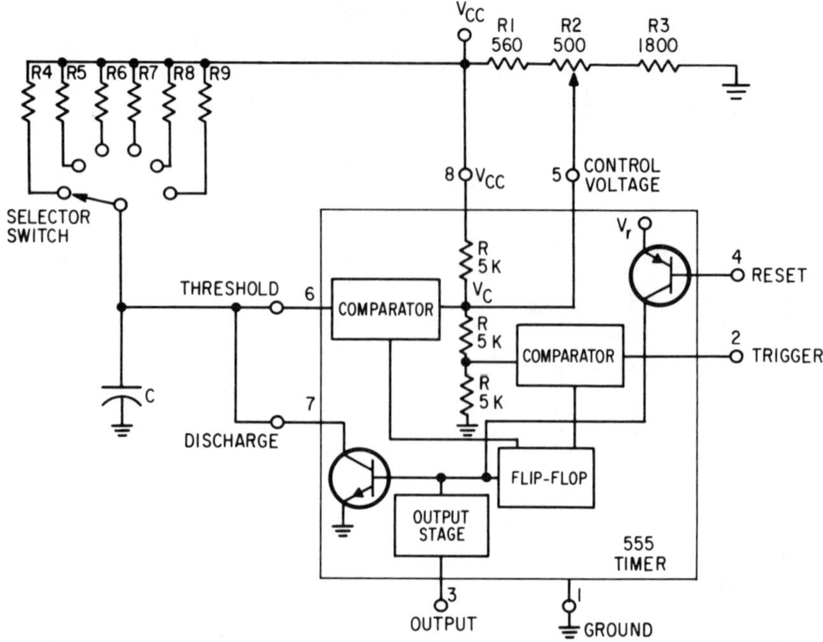

Fig. 2-10. Compensated monostable.

The solution in this case would be to use the control-voltage terminal (pin 5) to control the reference voltage of the comparator. Instead of connecting pin 5 to ground through a capacitor as usual, it is connected to trimmer resistor R2 (Fig. 2-10). By adjusting R2, it is now possible to control the reference voltage of the comparator, hence the threshold voltage at which the charging of the capacitor is terminated. This in turn allows the time delay to be varied to compensate for the combined tolerances of the timing capacitor and the charging resistor. The approximate effect on the time delay of these tolerances is given by the expression: $t = RC(1 \pm \Delta)\ln V_{cc}/(V_{cc} - V_c)$, where Δ is the worst case sum of the timing capacitor and resistor tolerances and V_c is the threshold voltage. For example, with a capacitor tolerance of $\pm 10\%$ and a resistor tolerance of $\pm 1\%$, Δ is 0.11. For a 555 in the monostable mode, the delay, or output pulse width, can also be expressed by: $t = 1.1RC$. Setting the two equations equal to each other results in: $1.1RC = RC(1 \pm \Delta)\ln V_{cc}/(V_{cc} - V_c)$.

By simplifying this expression, it is possible to get an expression that can be used to calculate the range of V_c required to compensate for the tolerances in the timing components. This equation is: $V_c = V_{cc}(1 - e^{-1.1/(1 \pm \Delta)})$.

Using this formula, the maximum and minimum values for V_c corresponding to the tolerance variations of the timing components can be calculated. For example, with a V_{cc} of 5 V, a $\pm 10\%$ capacitor tolerance, and a $\pm 1\%$ resistor tolerance, the V_c range required for compensation is 3.14–3.54 V.

Since we already know that the 555's internal voltage divider resistors are all 5-kΩ resistors (see p. 1), it is now possible to compute the values of R1, R2, and R3 after first choosing the desired current through this resistor string. In this example, using closest standard values and a 1.7 – mA current, R1 = 560 Ω, R2 = 500 Ω, and R3 = 1,800 Ω.

It should be noted that when multiple timing resistors are used, their tolerances will determine the accuracy of the time delay. In that case, it should be only the tolerance of the capacitor. In addition, control voltage V_c also affects the value of the trigger voltage. Normally it is 1.67 V. However, in the example given, this value will vary between 1.57 and 1.77 V, depending on the setting of trimmer R2. This usually will not cause any problems.

3.1   the basic astable

The astable multivibrator, also called a pulse generator or an oscillator, is a circuit that produces a series of pulses as its output. The circuit of a 555 connected as an astable is shown in Fig. 3-1, and a curve to aid in the selection of the timing components is shown in Fig. 3-2. The maximum frequency of oscillation can be as high as 300 kHz.

To get the 555 to operate in the astable mode, it is necessary to continuously retrigger it. The simplest way to do this is to connect the trigger input (pin 2) to the threshold input (pin 6). In addition, the timing resistor is now split into two separate resistors (R1 and R2) with their junction point connected to the discharge terminal (pin 7).

When power is applied to the circuit, the trigger and threshold inputs are both below $\frac{1}{3}V_{cc}$, the timing capacitor is uncharged, and the output is high. The output stays high for a period of time determined by: $t_1 = 1.1(R1 + R2)C$.

At the end of this period of time, the voltage on the timing capacitor will have reached $\frac{2}{3}V_{cc}$, the upper comparator in the 555 will trigger the internal flip-flop, and the capacitor will begin to discharge through resistor R2. The time it takes for the capacitor to discharge to $\frac{1}{3}V_{cc}$ is t_2 and is determined by: $t_2 = 0.693R2C$. As the value of the voltage on the discharging capacitor reaches $\frac{1}{3}V_{cc}$, the timer retriggers itself and again starts to charge up to $\frac{2}{3}V_{cc}$. This time, the time required to reach $\frac{2}{3}V_{cc}$ is less than the previous time because the charging cycle is not starting from 0 V, but from $\frac{1}{3}V_{cc}$. The time is given by: $t_3 = 0.693(R1 + R2)$. All subsequent charge cycles have the charging time.

Thus, the total time required to complete one charge/discharge cycle is given by $t = 0.693(R1 + 2R2)C$. This equation holds true for all conditions except the first cycle, which will have a slightly longer time due to the need to charge from zero. As in the monostable mode, the charge and discharge times and therefore the frequency are independent of the supply voltage.

Fig. 3-1. Astable multivibrator.

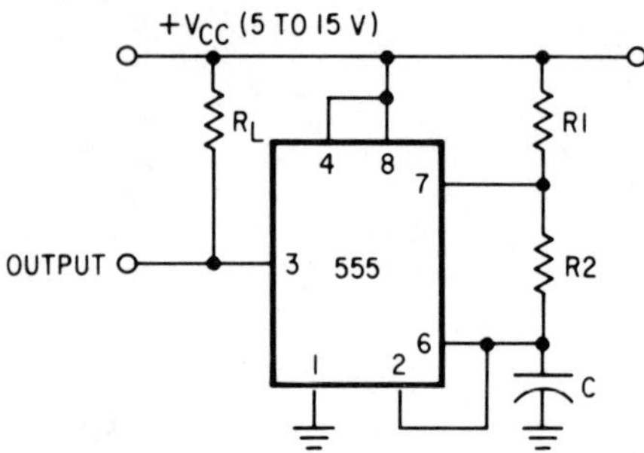

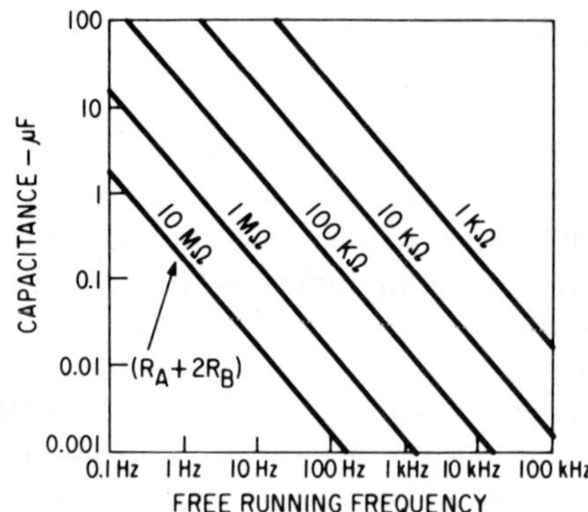

Fig. 3-2. Astable design curves.

The frequency of oscillation can be determined by inverting the equation for total cycle time. This leads to: $f = 1.44/(R1 + 2R2)C$.

Since the timing capacitor charges through R1 and R2, but only discharges through R2, the duty cycle (the "on" time divided by the total cycle time) may be precisely set by the ratio of R1 to R2. As R2 decreases, the duty cycle approaches 100%, and as it increases the duty cycle approaches (but never reaches) 50%. Duty cycle is determined by the following formula: $D = R2/(R1 + 2R2)$.

3.2   extended duty-cycle astable*

By adding a single diode to the basic astable circuit, the duty-cycle range of the astable can be almost doubled. Use of the diode makes the charging time constant of capacitor C independent of the discharge time constant (Fig. 3-3).

Connecting the diode between the threshold terminal (pin 6) and the discharge terminal (pin 7) makes it possible to achieve a duty cycle lower than the 50% minimum of conventional astables. Because of the diode's presence, the timing capacitor can now only charge through

* James, T. W., "Single Diode Extends Duty-Cycle Range of Astable Circuit Built with Timer IC," *Electronic Design*, March 1, 1973, p. 80.

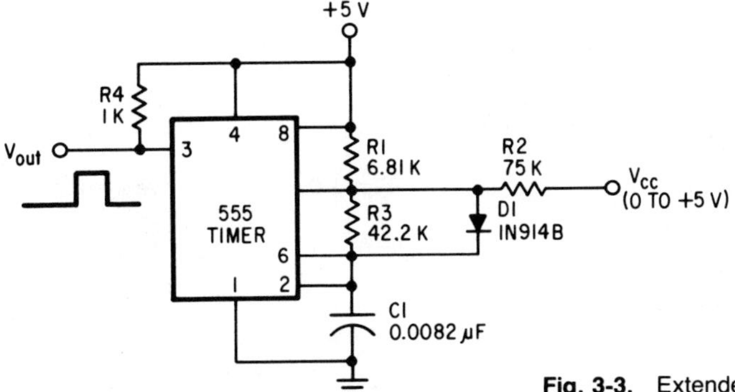

Fig. 3-3. Extended duty-cycle astable.

resistor R1. The reason for this is that during the charge cycle R2 is shorted by the diode (D1). During the discharge cycle only R2 is involved, as is the case for conventional astables.

The total cycle time for this circuit is given by the expression, $t = 0.693(R1 + R2)C$, and the duty cycle is given by, $D = R1/(R1 + R2)$. The frequency of oscillation can be calculated from, $f = 1.44/(R1 + R2)C$.

Although the addition of D1 causes the capacitor to charge through one resistor and discharge through another, this separation is not completely independent. The forward voltage drop across D1 prevents this. Complete independence can be achieved, however, by adding an additional diode to the circuit. This diode (D2) is added in series with R2 in a direction opposite to that of D1. All timing equations remain the same.

3.3   cheap astable

An inexpensive astable multivibrator can be built using the 555 timer, one resistor, and one capacitor (Fig. 3-4). In operation, the output (pin 3) is initially high and the timing capacitor starts charging exponentially with a time constant of RC. When the voltage across the

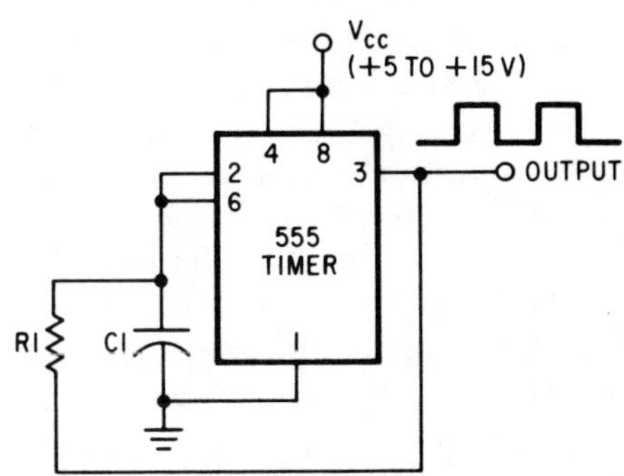

Fig. 3-4.   Cheap astable.

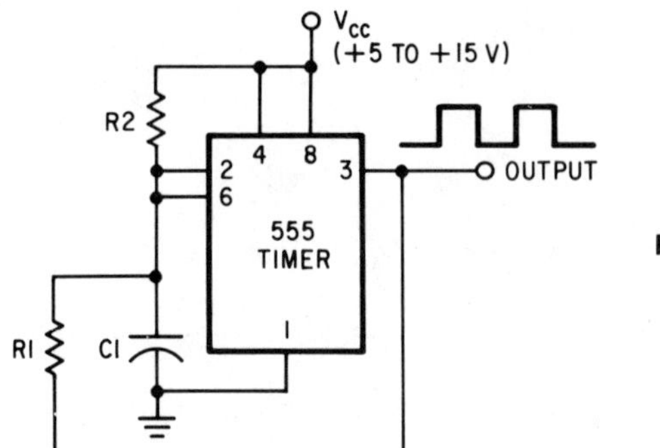

Fig. 3-5.   50% duty-cycle astable.

capacitor reaches the upper threshold ($\frac{2}{3}V_{cc}$) the internal comparator triggers the internal flip-flop and the output goes low (close to 0 V). The capacitor then starts to discharge exponentially with a time constant of RC. When the voltage across the capacitor reaches the lower threshold limit ($\frac{1}{3}V_{cc}$), the internal lower comparator triggers the internal flip-flop and the output goes high again, producing a series of pulses on the output. The cycle time is approximately 2(0.693)RC and the frequency of oscillation about 0.722/RC.

3.4   astable with 50% duty cycle*

Although the previous circuit produces an output pulse waveform, it should really produce a square wave: a pulse waveform with a 50% duty cycle. The reason for this is that the capacitor is charging and discharging through the same resistor (R1). However, since the voltage difference between the upper level of the output and the upper threshold are not exactly equal to the voltage difference between the lower threshold and the lower level of the output, the output remains in the high-level state longer than it stays in the low-level state, producing an asymmetric waveform.

This situation can be easily corrected by adding an additional resistor (R2) to the circuit (Fig. 3-5). This extra resistor is added between the positive supply rail (V_{cc}) and the junction of pins 2 and 6, the resistor, and the capacitor. This resistor makes it possible for the timing capacitor to charge up to the full supply voltage and thus makes the voltage difference between the upper level of the output and the upper threshold equal to the voltage difference between the lower threshold and the lower level of the output.

For any given values of R1 and V_{cc}, the value of R2 that produces a 50% duty cycle is easily determined experimentally. Once this value has been determined, the frequency may be varied without altering the duty cycle by changing the value of the timing capacitor.

* Hofheimer, R., "One Extra Resistor Gives 555 Timer 50% Duty Cycle," *EDN*, March 5, 1974, p. 74.

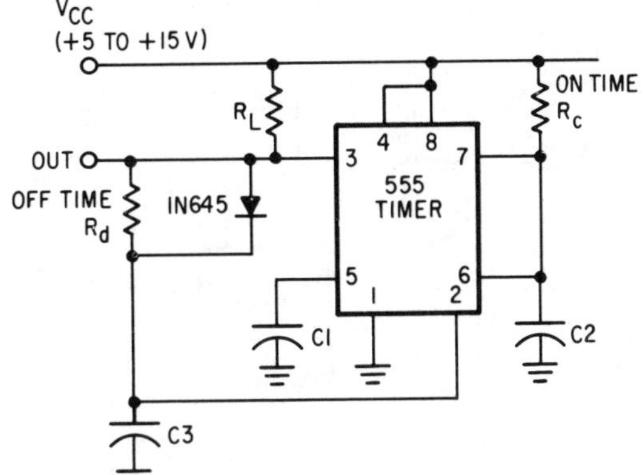

Fig. 3-6.   Improved astable.

3.5   improved astable*

By making a few alterations in the basic astable circuit and adding a few more components, an extremely versatile oscillator with totally independent "on" and "off" times can be produced. The modified circuit employs a second RC network that is controlled by the output terminal of the timer (Fig. 3-6).

During the charging time of C2 through R_c, the output voltage is held very close to V_{cc}. When C2 reaches $\frac{2}{3}V_{cc}$, the internal threshold detector clamps C2 and forces the output to a low state. Now, C3, which has been held at approximately V_{cc} by the output terminal, begins discharging through R_d. When the voltage on C3 discharges to $\frac{1}{3}V_{cc}$, the circuit triggers and the cycle repeats. The time constant for each part of the cycle are: $t_1 = 1.1R_cC2$ and $t_2 = 1.1R_cC3$. If C2 and C3 are set equal to each other, then the duty cycle of the oscillator will be given by: $D = R_c/R_d$.

Since the free-running period is simply the sum of the two time constants, the frequency of oscillation can be calculated by using the following formula: $f = 0.9/(R_c + R_d)C$, where $C = C2 = C3$.

3.6   astable with variable duty cycle†

By adding two transistors and a few resistors to the astable configuration of the 555, a multivibrator with a constant timing period and a duty cycle that can be adjusted over an extremely wide range can be constructed.

As in previous circuits that modified the duty cycle of the astable, it is necessary to provide separate paths for charging and discharging the timing capacitor (Fig. 3-7).

* Carter, J. P., "Astable Operation of IC Timers Can Be Improved," *EDN*, June 20, 1973, p. 83.
† Hilsher, R. W., "Constant Period with Variable Duty Cycle Obtained from 555 Timer with Single Control," *Electronic Design*, July 5, 1975, p. 72.

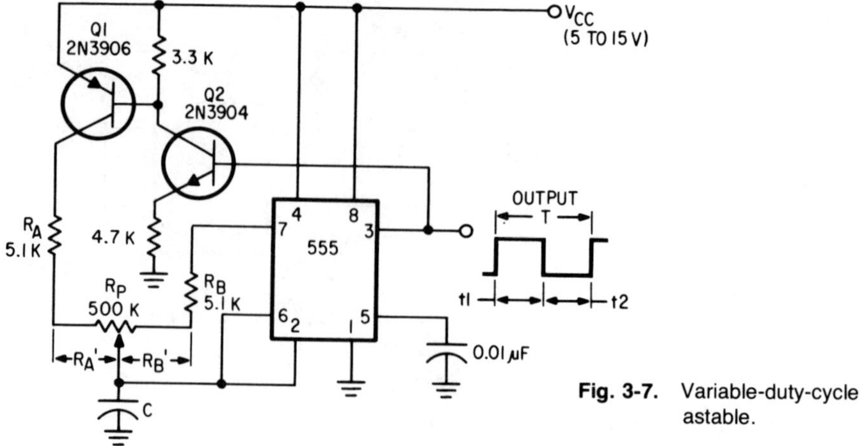

Fig. 3-7. Variable-duty-cycle astable.

Transistors Q1 and Q2 are both turned on when the output of the 555 (pin 3) is high. At this time, the timing capacitor is being charged by the current through Q1, resistor R_A, and the portion of the potentiometer designated $R_{A'}$. When the voltage across the capacitor reaches $\frac{2}{3}V_{cc}$, the output switches low.

At this point, the capacitor starts discharging toward the lower threshold limit of $\frac{1}{3}V_{cc}$ through $R_{B'}$, R_B, and the internal transistor of the 555. When the lower threshold is reached, the output switches high, and the timing cycle starts again.

The time during which the output is high is given by $t_1 = 0.693(R_A + R_{A'})C$. The time that the output is low is given by $t_2 = 0.693(R_{B'} + R_B)C$. The total cycle time is the sum of the two. Since $R_{A'} + R_{B'}$ is equal to R_p, the cycle time can be expressed as $t = (R_A + R_B + R_p)C$.

The duty cycle can now be changed by varying the potentiometer setting, but the total period and hence the frequency remain constant. The duty cycle is determined by the equation: $D = (R_A + R_{A'})/(R_A + R_{B'} + R_p)$. With the component values shown (Fig. 3-7), the duty cycle can be adjusted to any value between 1 and 99%.

3.7   crystal-stabilized astable*

A very stable multivibrator oscillator can be fabricated with the 555 by adding a crystal to the circuit (Fig. 3-8). The crystal is placed between the external RC series circuit and the timer's comparator terminals (pins 6 and 2). The charge/discharge paths for the timing capacitor remain the same as before, but the control signal to the comparators is now through the crystal. This forces the circuit to oscillate at the crystal frequency or one of its subharmonics.

* Althouse, J., "IC Timer, Stabilized by Crystal, Can Provide Subharmonic Frequencies," *Electronic Design*, Nov. 8, 1974, p. 148.

The values of R and C are selected so that with the crystal shorted out, oscillation will still be in the vicinity of the crystal frequency: f = 1.44/RC. But the R and C values can vary by 25% or more without the crystal oscillator's frequency being affected. However, the charge/discharge amplitude of C changes to accommodate the selected R and C values. This action keeps the frequency constant.

If the RC time constant is doubled, the oscillations shift to half the crystal frequency. Other changes produce corresponding subharmonics—$\frac{1}{3}$, $\frac{1}{4}$, $\frac{1}{5}$, etc.—of the crystal frequency.

The trimmer capacitor across the crystal allows precise but small adjustment of the frequency. And the resistor across the crystal provides a dc path for the comparator inputs to ensure that the oscillator will start when dc power is first applied.

3.8 high-accuracy astable

As was mentioned in the discussion of the basic astable circuit (p. 19), the first cycle of an astable is longer than all the succeeding ones because, when power is first applied, the timing capacitor must start charging up from 0 V, while on all subsequent cycles it starts charging from $\frac{1}{3}V_{cc}$.

In applications in which a low frequency clock signal is required, this extended cycle can cause problems. This problem can be quickly remedied by adding an external voltage divider to bias the timing capacitor at $\frac{1}{3}V_{cc}$ (Fig. 3-9). By making the external voltage divider

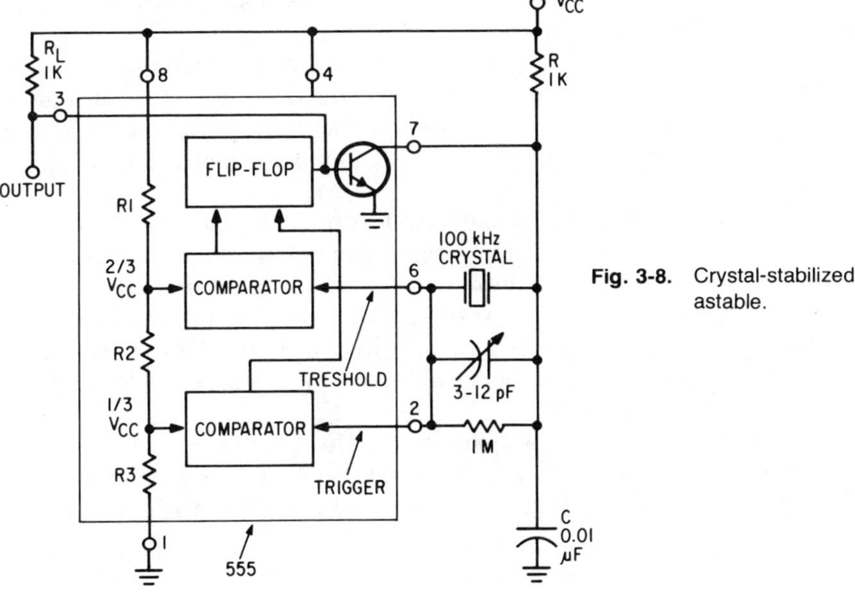

Fig. 3-8. Crystal-stabilized astable.

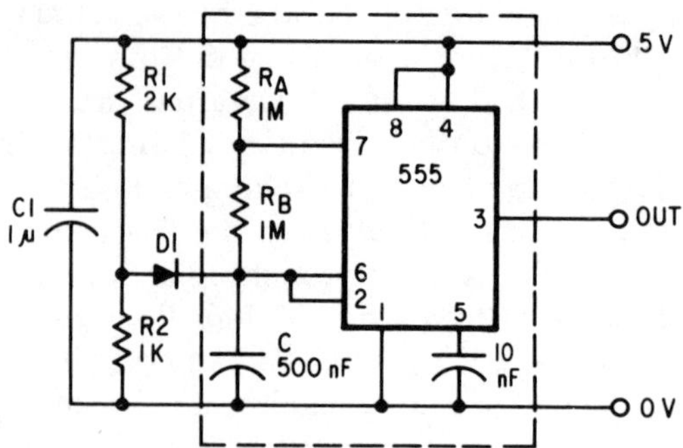

Fig. 3-9. High-accuracy astable.

resistors small compared with those of the conventional timer circuit, it is possible to charge the external timing capacitor up to ⅓V_{cc} before the first cycle has been completed. This will result in all cycles having the same length.

3.9   switch-selectable monostable/astable*

Using two RC networks and a switch, it is possible to come up with a multivibrator that can work in either the astable or monostable mode. When doing this, a problem arises because the periods of the two modes of operation differ. When the 555 is used as an oscillator, the timing capacitor normally charges from ⅓V_{cc} to ⅔V_{cc}. This provides an output period of 0.693RC. However, when the 555 is used in the monostable mode, or when strobed via the reset input, the capacitor must normally charge from 0 V to ⅔V_{cc}. The result is that a longer period of 1.1RC is produced.

The solid lines in Fig. 3-10 show the conventional circuit arrangement for the timer. The switch selects either the astable or monostable configuration. To equalize the time periods of the two modes of operation, either a resistor (R3) or a diode (both shown dotted) can be used.

A diode placed between the output (pin 3) and the control voltage input (pin 5) pulls down the pin 5 reference voltage to about 0.9 V each time the timer output goes low. Thus, the timing capacitor must now drop to approximately 0.45 V before the level at pin 2 can trigger another output pulse. The capacitor therefore starts to charge from near ground level in both the astable and monostable modes and the periods agree to within 5%.

The advantage of the diode method is that no computations or high-accuracy components are required to provide close matching of the pulse widths. Also, single potentiometer control of the pulse widths

---

* Klinger, A. R., "Single Part Minimizes Differences in Monostable and Astable Periods of 555," *Electronic Design*, July 19, 1974.

is still possible. However, the lower threshold and therefore the pulse width depend on the diode's offset and drift characteristics.

In a second method, resistor R3 forces the monostable period to approach that of the astable when it prevents the timing capacitor from discharging completely. Careful adjustment of the voltage divider formed by R1 and R3 permits the timing capacitor voltage to drop only far enough to trigger a new pulse. The timing capacitor starts to charge from about two-thirds of the supply voltage in both the monostable and astable modes. The advantage of the resistor method is that the periods of the two modes are governed by the adjustment of R1 and R3. Thus, the periods can be set very close to each other and a bypass capacitor to ground can be placed on pin 5, as is done normally. Also, the resistor method does not introduce the temperature drift of a diode, and the match of pulse widths tends to remain constant with supply-voltage variations.

One disadvantage is that the value of R1 cannot be varied to control pulse period without adjustments to R3 also. In addition, careful consideration of the tolerances of both internal and external voltage-divider resistances is required to attain close pulse-period matching. A cursory analysis shows that 5% resistors of 4.7 and 1.5 kΩ for R1 and R3 yield pulse periods matched to about 20%.

One percent resistors would allow considerably closer matching. The resistor method is best when high stability is required, or when it is desirable to bypass or to modulate pin 5, and when it is not necessary to have a continuously adjustable pulse width.

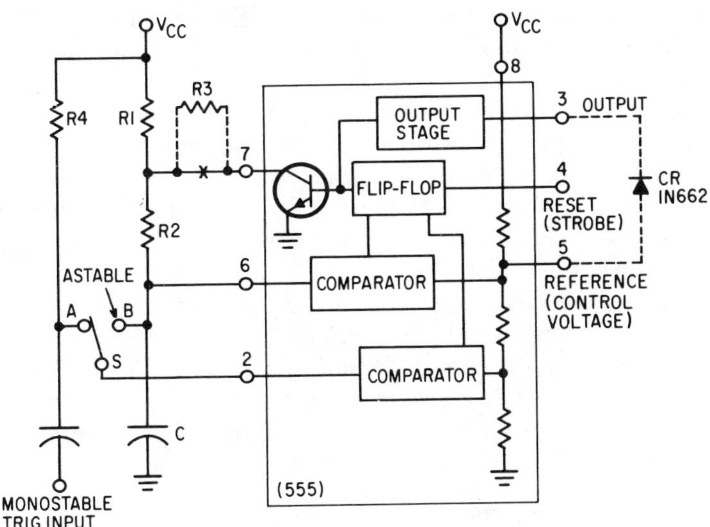

Fig. 3-10. Switch-selectable monostable/astable.

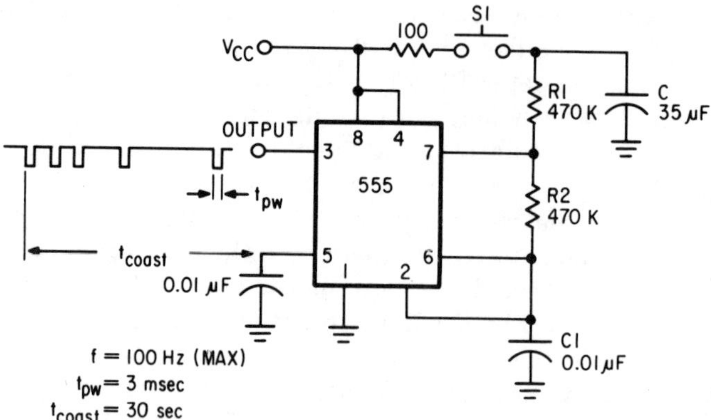

Fig. 3-11.   Diminishing-frequency astable.

3.10   diminishing-frequency astable*

For many electronic game applications, such as roulette, it is desirable to have an oscillator that starts oscillating at one frequency and gradually decreases in frequency until it stops. An oscillator that will coast to a stop like that while maintaining a constant output pulse width is seen in Fig. 3-11.

When switch S1 is closed, capacitor C charges to the supply voltage (V_{cc}) almost immediately, and the 555 oscillates at its highest frequency, which is determined by the standard astable formula: $f = 1.44/(R1 + 2R2)C$.

After the switch is released, capacitor C supplies voltage to the timing circuit. But as C discharges, the timing capacitor C1 takes longer and longer to reach its trip point of $\frac{2}{3}V_{cc}$. Thus, the oscillation frequency decreases.

The discharge time, t_{pw}, as C1 discharges to $\frac{1}{3}V_{cc}$ via R2 and the open collector output of the 555 (pin 7), remains constant, thus the output pulse width remains constant. But the time between pulses increases until capacitor C can no longer supply enough charge to recharge C1 to $\frac{2}{3}V_{cc}$. When the oscillations cease, the output (pin 3) of the 555 remains high.

For small output-pulse widths (C1 small), C can be a relatively small electrolytic capacitor. For example, if R1 and R2 are each 470 kΩ and C1 is 0.01 µF, then C need be only 35 µF for a 30-sec coast time. The frequency then starts at 100 Hz with a 3-msec-wide, negative-going output pulse. It gradually coasts down to zero frequency, while maintaining the 3-msec pulse width over the total 30-sec period.

* Kraengel, W. D., Jr., "Timer Pulses Coasting to a Stop Heighten Electronic Game Realism," *Electronic Design*, 1977.

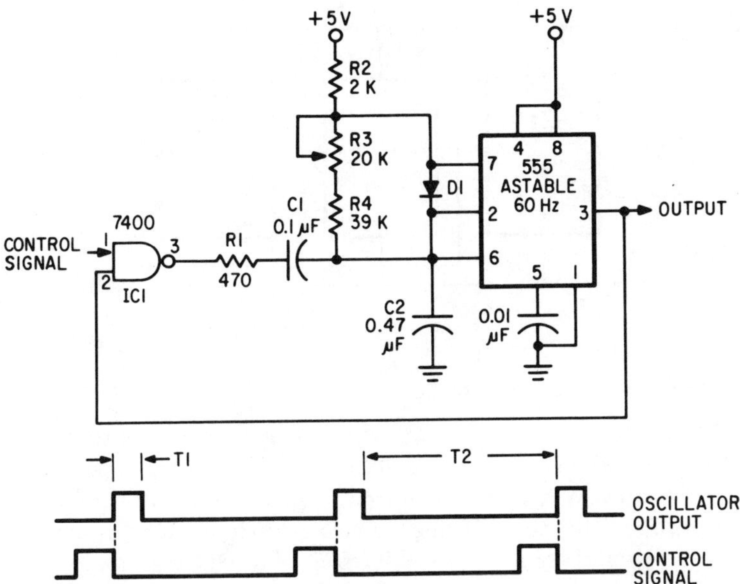

Fig. 3-12.   Frequency-locked astable.

3.11   frequency-locked astable*

By adding a two-input NAND gate to the standard astable multivibrator circuit, it is possible to build a locked oscillator (Fig. 3-12). The circuit will run free with no control input and will lock onto a control-input signal if the signal is near enough to the frequency of the free-running astable.

The timer, connected as an astable, uses resistors R2, R3, and R4, capacitor C2, and diode D1. During the time that C2 is being charged (pin 3 is high), R2 plus the resistance of the forward-biased diode D1 determine the time constant t_1. During the discharge of C2, diode D1 is reverse biased and the discharge time constant t_2 is determined by resistors R3 and R4.

With no control-input signal (low input to the NAND gate), the output of the NAND gate is high and the astable runs free. With a control signal present, the NAND gate's output goes low, discharging C2 when both the control signal and astable output are high. The astable then starts a new timing cycle.

The astable free-running frequency should be near the control-signal frequency. Changing the value of C1 controls the lock range. The oscillator will also lock onto harmonics of the control signal.

* Saunders, L., "Locked Oscillator Uses a 555 Timer," *EDN*, June 20, 1975, p. 114.

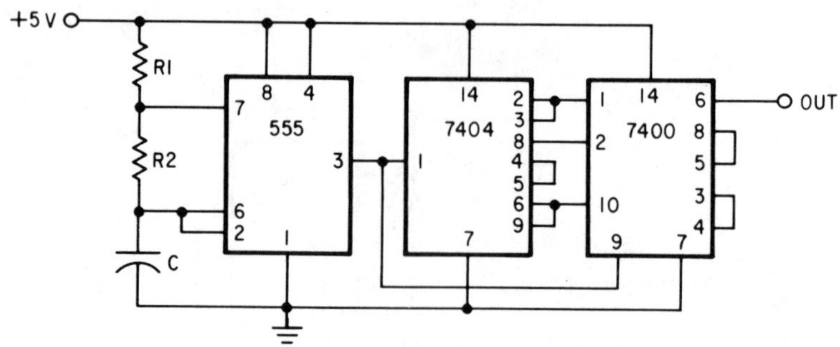

Fig. 3-13.   Increased-range astable.

3.12   increased-range astable

The normal maximum frequency of operation for the 555 timer is limited to 300 kHz (maybe less, depending on the manufacturer). Sometimes, however, it is necessary to have an oscillator with a frequency higher than that. By using two additional ICs, along with the conventional astable, it is possible to double the frequency available from the astable.

The heart of the circuit is the basic astable, the output of which is fed to the frequency-doubler circuit, which is made up of IC2 and IC3 (see Fig. 3-13).

The frequency-doubler circuit passes the output of the astable circuit through a series of logic gates in such a way that the propagation delay of the gates (the time it takes for the gate to switch from one logic level to another) are used to generate two 60-nsec-wide pulses for each incoming pulse, no matter what its frequency. The output frequency can be calculated from the following formula: $f = 2.88/(R1 + R2)C$.

The output waveform is not completely symmetrical because there is a 20-nsec difference in the time between successive pulses. But this is not noticeable at the frequencies that the 555 operates. The doubler circuit could be used up to frequencies as high as 10 MHz, in which case the asymmetry would be plainly noticeable. If desired, several doubler stages can be connected in series to achieve even higher frequencies.

3.13   complementary astable

If two monostable multivibrators are cross coupled (the output of one connected to the input of the other and vice versa) a complementary astable multivibrator can be built in which one monostable acts as a source for current, while the other unit acts as a sink for current (Fig. 3-14).

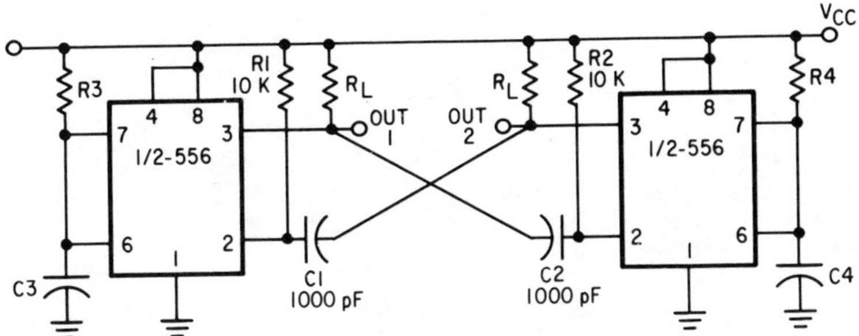

Fig. 3-14.   Complementary astable.

In operation, the output of each monostable is differentiated by the R1C1 and R2C2 differentiating networks. The R1C1 and R2C2 time constants should be at least 10 times smaller than the R3C3 and R4C4 time constants, which determine the "on" times of each monostable and, hence, the frequency of operation.

The timing period of each monostable is independently adjustable and calculated according to the standard monostable formula, $t = 1.1RC$, where R and C would be R3C3 and R4C4, respectively.

For symmetrical operation, where each monostable is on for the same amount of time, the total period would be equal to $2.2RC$ and the frequency of oscillation would be: $f = 0.455/RC$. For asymmetrical operation, the formula would be: $f = 0.91/(R3C3 + R4C4)$.

4.1   line driver*

With a little careful thought, the 555 timer can be made to do things it wasn't originally designed to do. An example of this is its use as a line driver (Fig. 4-1).

In this circuit, the threshold and discharge terminals (pins 6 and 7) are connected together and pulled up to the supply voltage bus by a 4.7-kΩ resistor. At the same time, the trigger input (pin 2) is grounded. Grounding the trigger input causes the voltage applied to the internal voltage comparator to drop below $\frac{1}{3}V_{cc}$. This causes the output to switch from a low state to a high state, where it will be maintained. As long as the output is high, the discharge transistor inside the timer is turned on and pin 7 is effectively shorted to ground. Since pin 6 (threshold input) is connected to pin 7 also, it too is shorted to ground. Since the voltage on the threshold pin is less than $\frac{2}{3}V_{cc}$, the comparator connected to it does not reset the internal flip-flop and thus helps maintain the output in its high state.

If an input is now applied to the reset terminal (pin 4), the output of the 555 will follow the input. The reason is that the reset terminal is activated when a voltage level anywhere between 0 and 0.7 V is applied to the pin. When activated, the reset pin will force the output of the 555 (pin 3) to go low, no matter what the other inputs to the flip-flop are. Thus, as long as the input to pin 4 is high, the reset pin is inactive and the output stays high. But when a logic low is applied to pin 4, it resets the output and forces it low for as long as this signal is present at pin 4, and then returns to the high state.

4.2   inverting line driver

The line driver in circuit 4.1 can be converted to an inverting line driver by using the trigger terminal (pin 2) as the input and tying the reset input (pin 4) to the positive supply rail (Fig. 4-2).

* Gardner, M. R., "Line Drivers Made from 555 Timers Provide Inverted or Non-inverted Outputs," *Electronic Design*, Jan. 19, 1976, p. 86.

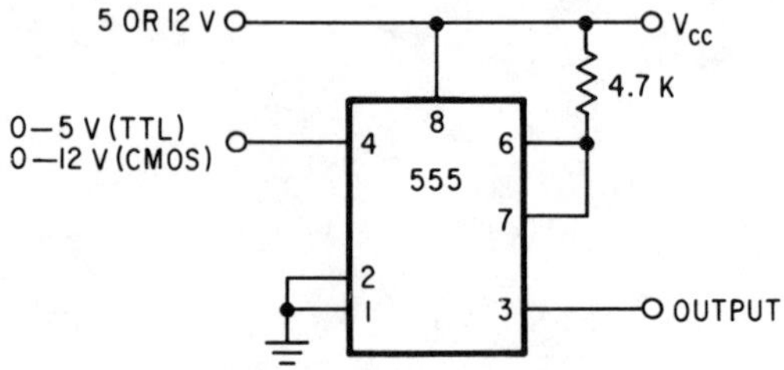

Fig. 4-1. Line driver.

In this case, the trigger input is not normally activated and the output is therefore low. If the output is low, the discharge transistor is off and the junction of pins 6 and 7 floats up to V_{cc}. Since this applies a voltage greater than $\frac{2}{3}V_{cc}$ to the threshold input (pin 6), the internal comparator is triggered and the flip-flop in the 555 acts to maintain the low state at the output (pin 3). Pin 4 (the reset) is inactivated by connecting it to the positive supply bus.

Now, when a negative-going pulse causes the voltage at this point to drop below $\frac{1}{3}V_{cc}$, the comparator to which this input is connected causes the internal flip-flop to change state, causing the output level to switch from low to high. When the input at pin 2 returns to its high state, the output returns to its low state, thus producing an inverting line driver.

Both this circuit and the previous one produce an output voltage swing that almost reaches each bus, and the driver can source or sink 200 mA. This drive capability is adequate for either single-ended or balanced lines that terminate in optical isolators.

In addition, since the 555 input in each of these circuits acts as a comparator, it can simultaneously serve as a level shifter. For example, with only a 5-V signal at the input, a 12-V swing can be applied to the line.

Additional problems encountered with these two circuits include a limited speed of 100–300 kHz (depending on the manufacturer of the device) and an inability to work with input levels lower than $\frac{1}{3}V_{cc}$ for the inverting line driver. This can be remedied somewhat, however, by using a pullup resistor and connecting it between pin 2 and V_{cc}.

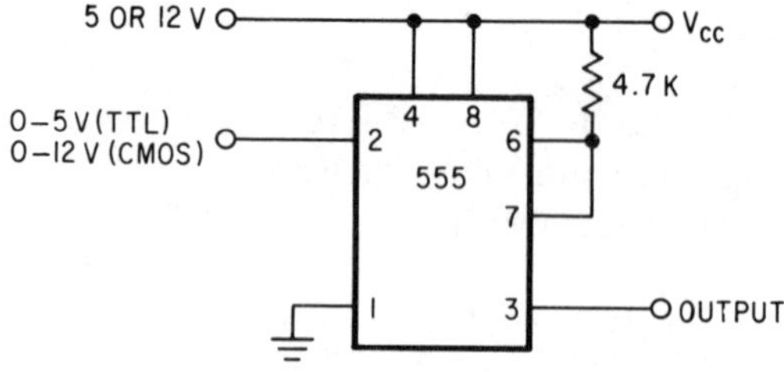

Fig. 4-2. Inverting line driver.

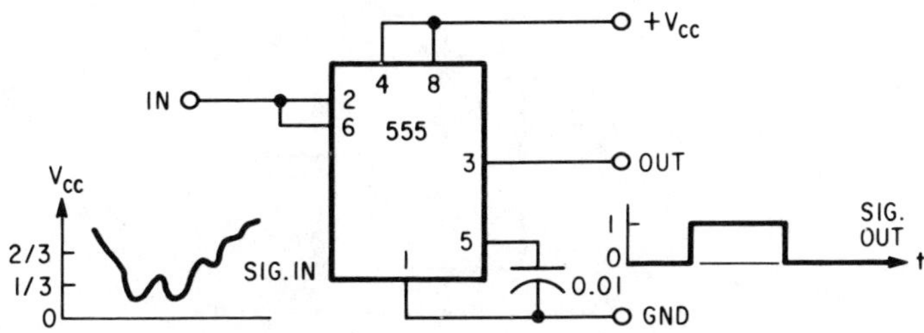

Fig. 4-3.   Simple Schmitt trigger.

4.3   simple Schmitt trigger

The 555 IC timer can be easily connected to operate as a Schmitt trigger. And the output can drive up to 6 TTL loads. A Schmitt trigger is a circuit that does nothing until a certain input level is reached, at which point its output goes high. The output then stays high until a different level of the input signal is reached. At that point, the output of the Schmitt trigger goes low again. Such a circuit could be used to convert a half-wave rectified sine wave to a square wave (useful for digital clock circuits).

To use the 555 as a Schmitt trigger (Fig. 4-3) only requires that the trigger terminal (pin 2) and the threshold terminal (pin 6) be tied together and used as the input. Now, when an input signal applied to these two points drops below $\frac{1}{3} V_{cc}$, the comparator to which this pin is connected causes the internal flip-flop of the 555 to change state. This forces the output terminal (pin 3) to go high. The timer will remain in this new state until another signal resets the flip-flop.

When the input signal rises again to a value greater than $\frac{2}{3} V_{cc}$, a second internal comparator is triggered and resets the flip-flop, causing the output to return to its low state. The difference between the two trigger levels of the Schmitt trigger is called the hysteresis. With a 5-V supply, hysteresis in this circuit is 1.6 V.

4.4   adjustable Schmitt trigger*

One disadvantage of the previous circuit is that the threshold levels are fixed at $\frac{1}{3}$ and $\frac{2}{3} V_{cc}$. But, by adding two resistors (Fig. 4-4), it is possible to reduce the hysteresis from the 1.6 V of the previous circuit to as little as 0.5 V.

The lower trip point in this circuit occurs when $V_{in} R2/(R1 + R2) = \frac{1}{3} V_{cc}$. To allow for internal voltage variations, the input voltage should never be allowed to drop below $\frac{1}{3} V_{cc} + 0.2$ V. The upper trip point will occur as usual at $\frac{2}{3} V_{cc}$.

* Hoven, D.,"IC Timer Makes Adjustable Schmitt Trigger," *EDN*, May 5, 1973, p. 73.

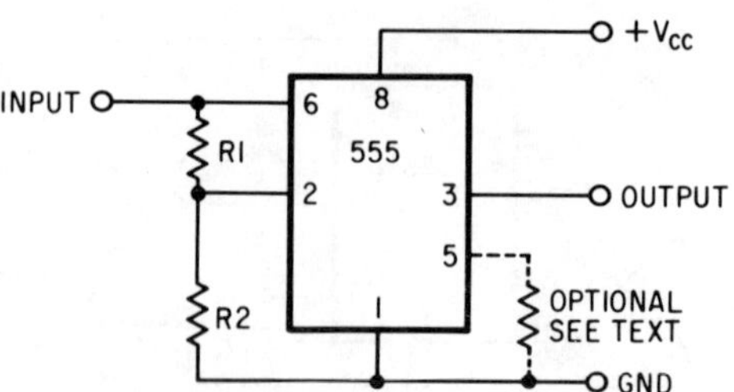

Fig. 4-4.   Adjustable Schmitt trigger.

A further modification can be made to the hysteresis by lowering the voltage at pin 5 by connecting a resistor to ground. Pin 5 happens to be an input for the upper comparator of the 555, and by placing a resistor to ground at this point, the threshold of the comparator is changed.

The upper trip point thus occurs at whatever the voltage is at pin 5. And the lower trip point occurs when: $V_{in}R2/(R1 + R2) = V_{pin\ 5} + 0.2\ V$.

4.5   another adjustable Schmitt trigger

Another way of varying the hysteresis of a Schmitt trigger is to do it with zener diodes and a transistor as in Fig. 4-5. The circuit operates as follows: If the input signal equals zero, the output goes high, causing the transistor Q1 to turn on. Thus, the controlled voltage input (pin 5) receives a voltage equal to ($V_{CE(sat)} + V_{z1}$). This is the upper threshold. V_{z1} is the voltage rating of the first zener diode and $V_{CE(sat)}$ is the collector-to-emitter saturation voltage of Q1.

As the input voltage increases to $\frac{2}{3}V_{cc}$, the output goes low and Q1 turns off. Thus, the voltage at terminal 5 equals $V_{z1} + V_{z2}$. Since about half of this voltage hits the second comparator, the lower threshold point will approximately equal $(V_{z1} + V_{z2})/2$. By choosing various

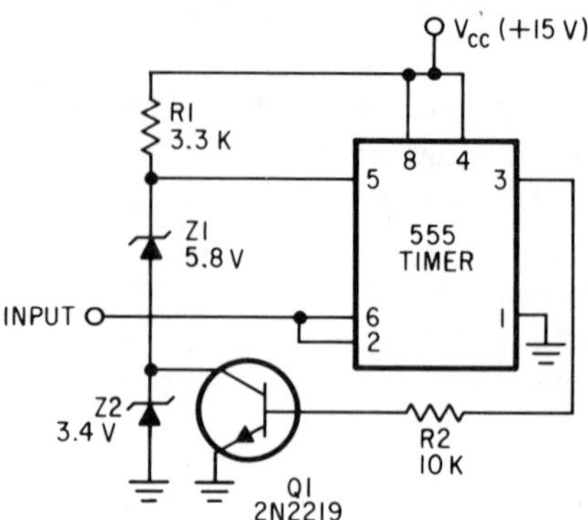

Fig. 4-5.   Another adjustable Schmitt trigger.

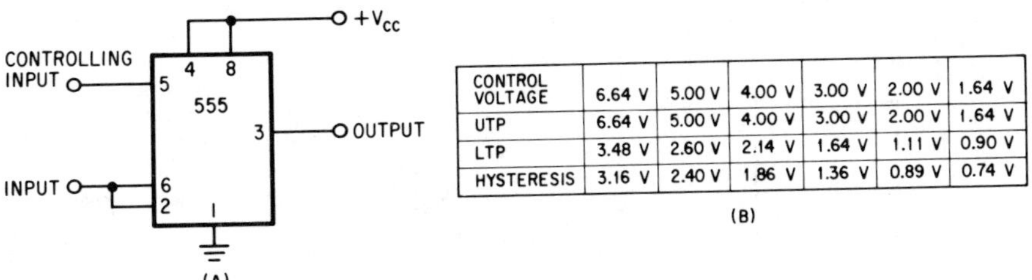

CONTROL VOLTAGE	6.64 V	5.00 V	4.00 V	3.00 V	2.00 V	1.64 V
UTP	6.64 V	5.00 V	4.00 V	3.00 V	2.00 V	1.64 V
LTP	3.48 V	2.60 V	2.14 V	1.64 V	1.11 V	0.90 V
HYSTERESIS	3.16 V	2.40 V	1.86 V	1.36 V	0.89 V	0.74 V

(B)

Fig. 4-6. Controlled-input Schmitt trigger.

zeners, it is easy to pick any desired threshold points. For the circuit in Fig. 4-5, the lower threshold is 4.7 V and the upper is 5.8 V.

4.6 controlled-input Schmitt trigger*

Sometimes it is desirable to have a Schmitt trigger whose threshold levels, and hence hysteresis, is adjustable by changing a voltage level. This comes in handy in medical equipment, in which input signals may vary widely in amplitude.

Making such a trigger is very simple. Instead of connecting pin 5 to ground through a capacitor (as was done in circuit 25), use pin 5 as the controlled voltage input (Fig. 4.6A). In a circuit like this, the upper threshold point is fixed at the level of the applied control voltage, while the lower threshold point is a little more than half this value (Fig. 4.6B).

4.7 bistable flip-flop†

The 555 can be used as a bistable (two stable states) flip-flop by not using an RC timing network and by tying the threshold input (pin 6) low (Fig. 4-7). The output state of the flip-flop configuration is determined by the trigger and reset inputs. These inputs are pulled high by resistors R1 and R2, then pulled low through either a switch or a TTL logic level "zero." A logic "zero" on the reset terminal (pin 4) will drive the output of the 555 low until the reset goes high and the trigger goes low.

Due to this bistable action, contact bounce of a mechanical switch will not cause erroneous switching of the output.

* Lalitha, M. K., and Chetty, P. R. K., "Variable Threshold Schmitt Trigger Uses 555 Timer," *EDN*, Sept. 20, 1976, p. 112.
† Gephart, R. L., "Mini-DIP Bistable Flip-Flop Sinks or Sources 200 mA," *EDN*, Oct. 5, 1974, p. 76.

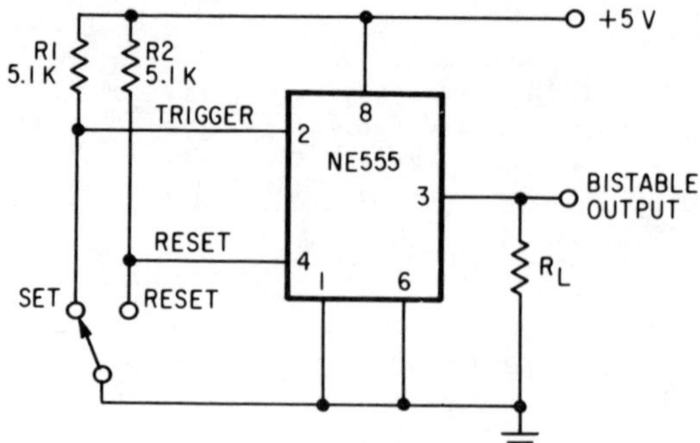

TRUTH TABLE

TRIGGER (PIN 2)	RESET (PIN 4)	OUT
↓	HIGH	↑
↑	HIGH	HIGH
HIGH	↓	↓
HIGH	↑	LOW

Fig. 4-7.   Bistable flip-flop.

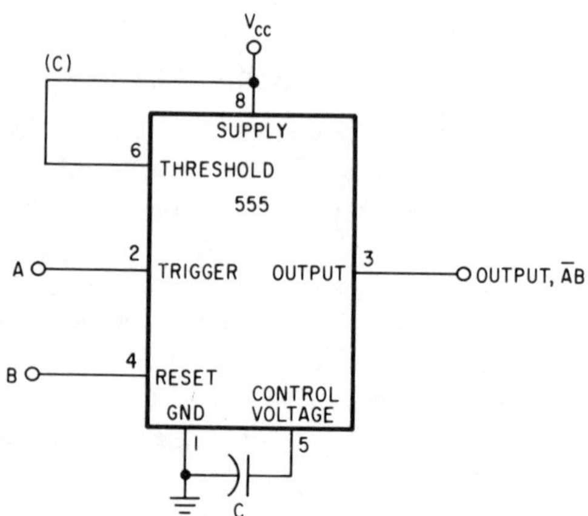

Fig. 4-8.   NOT AND gate.

4.8   NOT AND gate*

By tying the threshold terminal (pin 6) up to the positive supply rail and using the trigger (pin 2) as the A input and the reset (pin 4) as the B input, it is possible to use the 555 to perform the logic function NOT AND ($\overline{AB}$) as can be seen in Fig. 4-8.

As long as B (the reset input) remains low, the output of the 555 remains low. When B goes high, the output of the timer goes high too. Now, if a logic high is applied to the A (trigger) input, it will produce an inverted output at pin 3, and vice versa if the A input is a logic low. Since the output of the 555 is unaffected by the A input when B is low, but gives an inversion of A when B is high, the $\overline{AB}$ function has been implemented.

* Murugesan, S., "Create a Versatile Logic Family with 555 Timers," *EDN*, Sept. 5, 1976, p. 108.

chapter
five

timer-based
instruments

5.1   cheap timer tester

A simple circuit that can be used to perform a functional test on 555 timers can be built by adding two light-emitting diodes (LEDs) and two resistors to the standard astable multivibrator configuration of the IC timer (Fig. 5-1).

The resistors and capacitor in the frequency-determining portion of the circuit are determined so that an output frequency of a little less than 1 Hz is produced. To do this, R1, R2, and C are chosen as 1 kΩ, 1 MΩ, and 4.7 μF, respectively. The output of this circuit will be close to a perfect square wave because the difference between the charging resistance and the discharging resistance is only about one-tenth of a percent.

The output of the timer is fed to two resistor-LED combinations in such a way that, if the timer is functional, it will alternately flash the two LEDs. When the output of the timer goes high, the voltage drop across the R3D1 combination is almost zero, because the anode side of the diode gets 9 V from the positive supply rail, while the cathode side of the diode gets almost 9 V from the output at pin 3. Thus, D1 will not light. It will only light when the output at pin 3 changes state to a low level.

The R4D2 combination, however, does light when the output of pin 3 is high because there is a potential difference between the cathode of D2 and pin 3 that is large enough to cause the LED to light. When the output at pin 3 reverses, however, the cathode of D2 is at ground potential and so is the output of pin 3. Thus, D2 indicates the presence of a high voltage on pin 3.

5.2   go/no go timer tester*

A slightly more sophisticated timer tester can be built with two IC timers, or a single dual timer device (Fig. 5-2).

* Predescu, J., "Tester Built for Less Than $10 Gives GO/NO GO Check of Timer ICs," *Electronic Design*, May 24, 1974, p. 106.

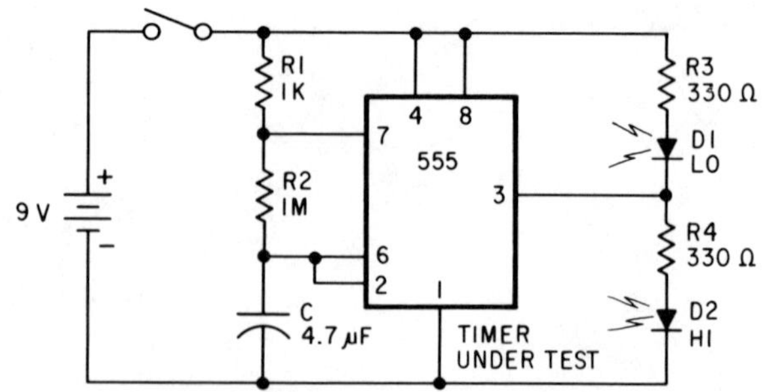

Fig. 5-1.   Circuit used to perform functional test on 555 timers.

The two timers determine the allowable accuracy for the timer IC under test. Potentiometers R1 and R2 permit ready adjustment for the desired range.

With power applied, all timers switch to the high state and begin their cycles. The output of IC1 inhibits the flip-flop for the interval T1. At T2, the output of IC2 goes low and inhibits any signal from the timer under test. the period between T1 and T2 is the time alloted for IC3, the timer under test, to complete its cycle and produce a low output. Only during this time can a high-to-low transition from IC3 trigger the flip-flop so that the D1 LED, which indicates a good IC, lights up. D2 lights up when the test is completed.

Although there can be a few milliseconds of contact bounce when S1 is first closed, thereby causing a delay in capacitor charging, the delay appears across all of the ICs. But since the ratio of delay times among all three timers is the same, the effect on test accuracy is nil.

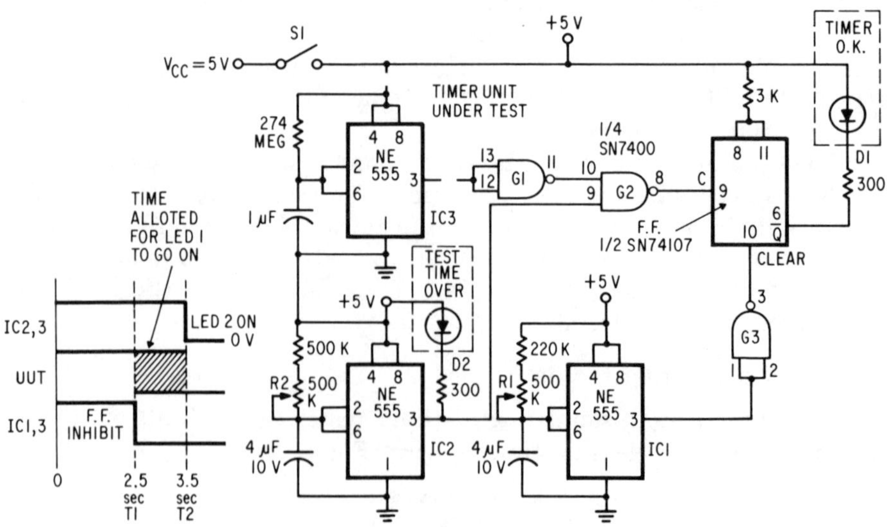

Fig. 5-2.   Go/no go timer tester.

5.3   multiple-trace oscilloscope display

Very often it is necessary to view two analog signals at the same time. If one has a multiplexing switch that converts a single trace scope to a dual trace then there's no problem. But if no such multiplexor is available, a simple timer circuit can be built that will label each waveform individually with its own distinctive pattern of long or short lines.

When a signal is fed into the circuit in Fig. 5-3, it can be converted from the continuous line signal that appears at the input, to a broken line equivalent. The length of the dashes in the broken line is determined by the frequency of the astable oscillator formed by the 555 IC.

The frequency of the astable should be at least three times larger than the input frequency of the analog signal; this will give a few long broken lines. As the frequency is increased, the lines will get shorter. The astable shown will allow an adjustment of frequency that varies from about 500 Hz to about 27 kHz. If a different frequency is desired, the standard astable formula, $f = 1.44/(R1 + 2R2)C$, can be used to determine the required components.

The method used to mark the waveform is very simple. The vertical position of the oscilloscope trace is determined by the sum of the voltages across resistors R3 and R4. A circuit formed by R4, C1, and D2 clamps the square-wave output of the 555 at 0 V (it passes the positive portion of the waveform while shunting the negative portion to ground). Thus, during one-half of the cycle, the full output voltage of the 555 is applied across R4, and during the other half, the voltage across R4 is almost zero.

Now if the output voltage of the 555 is several times larger than the amplitude of the signal to be displayed, and if the oscilloscope is set to display the input signal at full scale, when the input signal and

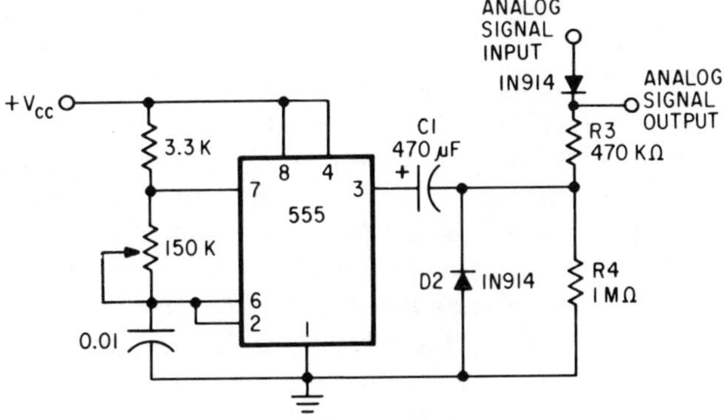

Fig. 5-3.   Multitrace oscilloscope display.

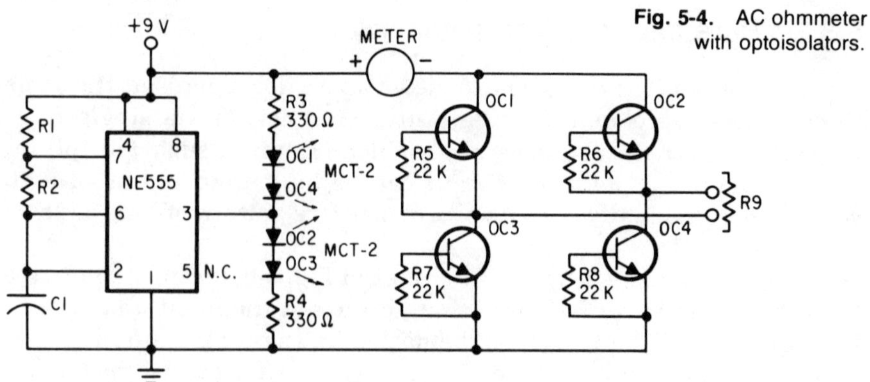

the output of the 555 are added together, as they are at point "A," during part of the cycle the scope trace will be displayed as normal, and during the other part it will be off the screen because the amplitude is too great.

At normal intensity settings, the square-wave switching waveform is not visible. C1 is selected such that the time constant R4C1 is large enough so that C1 will not discharge very much during the period of the input signal. D1 is used as an isolating diode to isolate the signal source from the 555 square-wave generator.

5.4   ac ohmmeter with optoisolators*

In the geophysical world, many things can be learned about soils and construction materials by measuring their electrical resistance. However, dc resistance readings are unreliable due to polarization and earth-current effects. An ac ohmmeter compensates for these errors, since the net current in the load is zero.

While conventional ac meters use bulky transformers and large capacitors in inverter-like circuits, this circuit eliminates them. Through a novel use of optically coupled isolators, ac current is developed in the load using a single battery (Fig. 5-4).

The 555 timer controls the output at a frequency determined by R1, R2, and C. The output of the timer will be high for a period of time determined by 0.69(R1 + R2)C and low for a period of time determined by 0.69(R2)C. Since in this application it is desirable that the output be as nearly symmetrical as possible, R1 should be much smaller than R2, but in no case less than about 1,000 Ω. The formula for the frequency of oscillation is the standard astable formula (see circuit 5.3).

---

* Beckwitt, D. J., "AC Ohmmeter Provides Novel Use for Opto-Isolators," *EDN*, July 5, 1974, p. 70.

The output switching matrix is controlled by the timer so that OC1 and OC4 are on for one-half cycle while OC2 and OC3 are on for the other half. The phototransistor pairs alternately saturate and switch the polarity of the voltage applied to the output terminals. Resistors R5 through R8 allow rapid turnoff of the phototransistors at high resistance values, preventing conduction overlap. the peak-to-peak output voltage is nearly equal to $2V_{cc}$. The output current is unaffected by frequency and duty cycle from 0 to 150 Hz.

Values for the resistors R3 through R8 will depend on the isolator used and the desired current output. R3 and R4 determine the current in the LED sections of the optoisolators according to the formula, $I = (V_{cc} - 2V_F)/R$, where V_F is the forward voltage drop of the LED.

In Fig. 5-4, Monsanto MCT-2 optoisolators were used. R3 and R4 are 330 Ω each, yielding a current of 20 mA with $V_{cc} = 9$ V. Resistors R5 through R8 are each 22 kΩ. Without the meter, this circuit can be used to develop ac from a single battery in many other applications.

5.5   IC timer resistance bridge*

A resistance bridge that makes use of two 555 timers operates without requiring the usual combination of a meter and an amplifier. Moreover, the circuit's sensitivity does not depend on the unknown resistance. And, since an LED is used for visual indication, there's no need to worry about shock isolation for a meter movement. Two possible applications for the bridge are a thermometer, where the unknown resistance could be a thermistor, or a photometer, where the unknown resistance could be a photoresistor (Fig. 5-5).

R_X is the unknown resistor in the bridge circuit. When the resistance of the dual potentiometer is increased, the brightness of the LED also increases, steadily. Then at a particular setting of the potentiometer, the LED's brightness is suddenly halved. The ratio of this particular setting on the potentiometer to R_X at which this decrease in brightness occurs is determined solely by the properties of the two IC timers.

The first timer (T1) operates in an astable mode and therefore is free-running. Its output, signal "A," is low for a period of time, $t1 = 0.69\,R_XC$ sec, and high for a period of time, $t2 = 0.69(R_X + R_{pot})C$ sec. The output from timer 1 is differentiated and then used to trigger the second timer (T2), which operates in a monostable mode. To simplify matters, the timing capacitors on both timers are set equal.

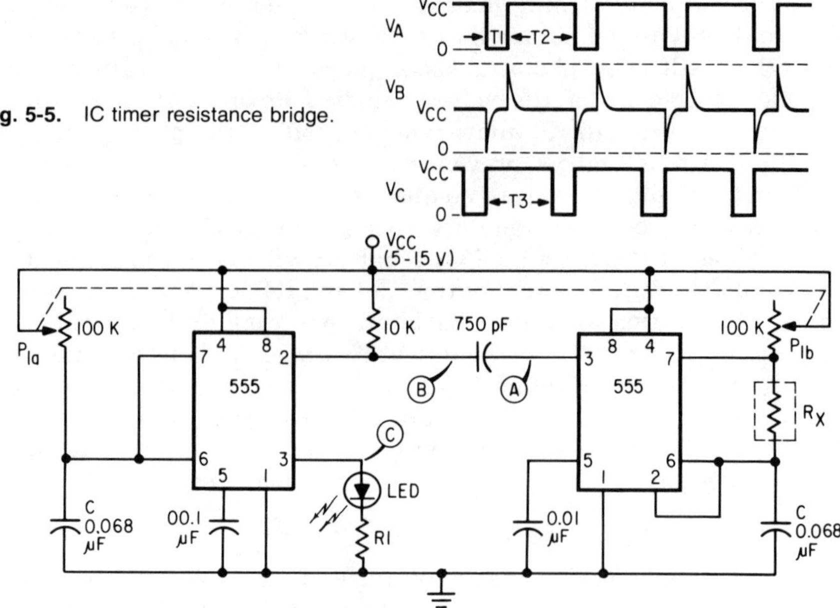

Fig. 5-5.   IC timer resistance bridge.

As the resistance of the potentiometer is increased, the periods of signals "A" and "B" become longer, and the on time of T2, which is $t3 = 1.1R_{pot}C$, starts to increase at a slightly faster rate. This means that the duty cycle of signal "C" is getting larger, and the LED will appear to glow brighter.

A closer look at the waveforms reveals that when period t3 is just slightly less than t1 + t2, the duty cycle of signal "C" is nearly 100%. But when t3 is slightly greater than t1 + t2, the duty cycle of signal "C" drops to 50% and, at the same time, the frequency of this signal decreases to half the frequency of signal "A." This happens because T2 locks out trigger pulses, while its output is still high and thus ignores all alternate negative-going spikes.

Further increases in the potentiometer resistance cause the duty cycle of the signal "C" to rise again slowly from 50% to a limiting value of 79.4%. The abrupt transistion from 100% to 50% occurs when the potentiometer resistance is equal to $3.406R_X$. This makes the calibration of the bridge intrinsically linear. Circuit performance is limited by the desired upper and lower operating frequencies and the width of the triggering pulses. For the component values shown, the circuit can operate over a fairly wide range of unknown resistance values, from 1 to 100 kΩ. The value selected for the LED's current-limiting resistor R1 depends on the supply voltage used.

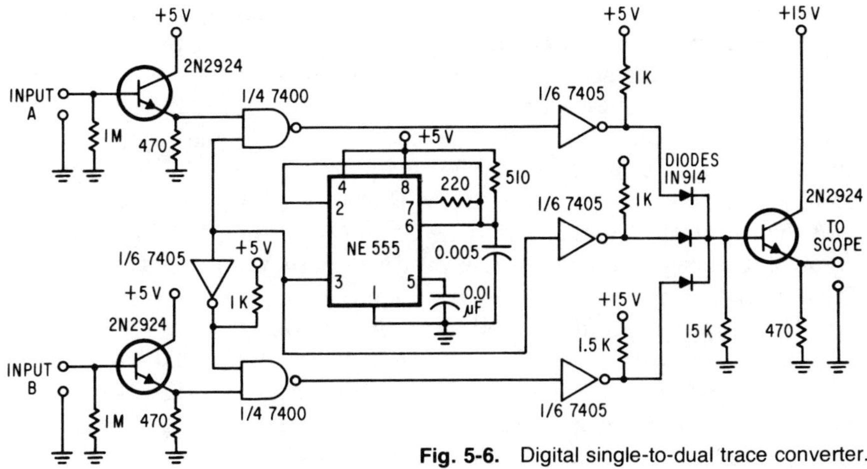

Fig. 5-6. Digital single-to-dual trace converter.

Fig. 5-7. Simplified dual-trace switch.

5.6 digital single-to-dual trace converter*

The display of two digital pulse trains on a single-trace scope can be accomplished with the aid of the circuit shown in Fig. 5-6. This converter will prove very useful in diagnosing logic circuit problems.

The circuit can be understood more easily with the help of the simplified circuit in Fig. 5-7. Gates G1, G2, and G3 are open collector AND gates. The collectors of these gates are connected to 5, 10, 15 V, respectively, via separate pullup resistors. Three diodes and resistor R_0 form a triple-input OR gate. The binary output levels of the OR gate are approximately 0–5, 10, or 15 V, as determined by the particular AND gate that is on. Gates G1 and G3 are switched alternately by the complementary outputs of the flip-flop. The rate of switching depends

* Tandon, V. B., "Circuit Converts Single-Trace Scope to Dual-Trace Display for Logic Signals," *Electronic Design*, April 12, 1975, p. 80.

on the clock frequency. And the resolution of the display improves with a fast clock. But the upper clock frequency is limited by the circuit components and the scope bandwidth characteristics.

When both inputs A and B are at zero, only gate G2 switches on and off and a square wave of approximately 10 V peak-to-peak amplitude appears at the output. The zero level of this square wave becomes the zero level for input A and the 10-V level becomes the zero level for input B. When input A is a logic one, about 5 V appears at the output. A logic one at input B produces a 15-V level at the output. Thus, digital signals at input A are displayed between 0 and 5 V, and signals at input B are displayed between 10 and 15 V.

In the actual circuit (Fig. 5-6), two 2N2924 transistors, hooked up as emitter-followers, are placed at the inputs A and B to reduce the load on the circuits under test: The 555 generates a 50% duty cycle square wave with a pulse width of about 2 μsec. Open-collector 7405 hex inverters complement the output of the timer and drive the 1N914 diodes.

5.7   simple logic-trace multiplier

Another multitrace oscilloscope adapter for logic signals can be built from only three ICs. This simpler circuit is shown in Fig. 5-8. In this circuit, a conventional astable 555 circuit is used to drive a flip-flop that provides a complementary output that is half the frequency of the 555 output signal.

The complementary outputs of the flip-flop (Q and $\overline{Q}$) alternately enable NAND gates G1 and G2, allowing first one channel and then the other to be displayed on the oscilloscope. Resistors R3 through R6 form a summing network. In order for the two input signals to be displayed in such a manner that they do not appear one on top of the other, R4 is used to provide a dc signal on top of which input 1 is added.

S1 makes it possible to provide a synch signal for either channel. The NAND gates wired as inverters (G3 and G4) invert the incoming

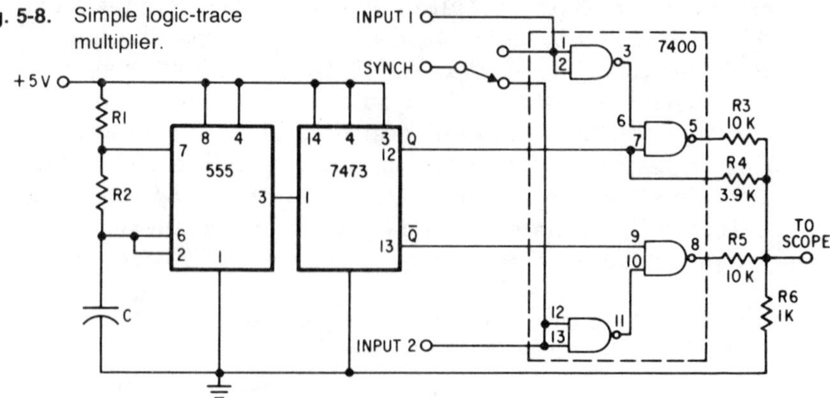

Fig. 5-8.   Simple logic-trace multiplier.

signal so that, after passing through gates G1 and G2, the output presented to the scope is the same as the original inputs. If these inverters were not used, the output would be inverted.

The frequency of the timer–flip-flop combination should be at least three times the input frequency and preferably more. It can be determined by the following equation: $f = 0.72/(R1 + 2R2)C$. This is just half the normal 555 astable frequency.

5.8   analog single-to-dual trace converter

By replacing the 7400 logic gates in the previous project with CMOS analog gates, it is possible to make a very simple analog single-to-dual trace scope converter. The trace doubler switches between the two input signals at high speed, permitting each to be displayed for half the time (Fig. 5-9).

To separate the two different signals, a dc offset voltage is added to the input signal. As in the previous circuit, an astable multivibrator is used to drive a flip-flop, which provides complementary switching signals to control the analog gates. In this case, however, the flip-flop is a CMOS unit 4027 and both the timer and the CMOS devices are referenced to -5 V instead of to ground. The analog switches are contained in a CMOS IC known as a quad bilateral switch and has the part number 4066 or 4016. Two switches are required for each channel. One switch gates the input signal on so that it can be displayed on the oscilloscope, while the other switch for that channel sets the dc level, which can be varied by a potentiometer. By varying the dc level, you vary the vertical position of the signal on the oscilloscope screen. The input signals to this trace doubler should be limited to an amplitude of less than 10 V peak-to-peak. Both channels are identical.

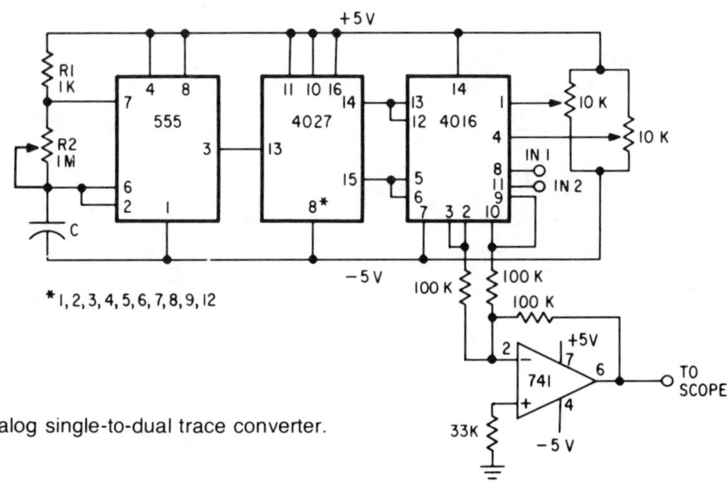

Fig. 5-9.   Analog single-to-dual trace converter.

After the signals pass through the electronic switches, they are fed to a unity-gain summing amplifier that is fabricated from a 741 operational amplifier. If you want to amplify or attenuate the combined signals coming out of the 741 op-amp, all that you have to do is change the value of R5. R5 and R3 and R4 determine the amplifier gain. If the ratio of R5 to R3 or R4 is 2, 5, 10, or ½, that's what the gain of the amplifier will be.

The frequency of the oscillator has been made adjustable because the unit will not display a correct trace if the oscillator frequency is an exact multiple of the frequency of the input signals.

5.9   oscilloscope calibrator

Another application of the CMOS analog switch and the 555 timer to the oscilloscope field is a scope calibrator. Most new scopes have a calibrated vertical axis, but some of the older ones don't. With this device (Fig. 5-10), you can measure the amplitude of the input signal by using the signal provided to adjust the gain on the Y amplifier of the scope so that one unit of height equals a certain amount of volts. Then count how many units high the unknown input signal is.

The heart of this calibrator is a 555 configured to operate in the astable mode. The oscillation frequency of the astable is determined by R1, R2, and C and has arbitrarily been set at about 1 kHz. The output of the timer, as in the previous project, is then fed to a CMOS flip-flop, which controls an analog gate IC. Only two of the four gates are used. One side of each gate is connected together and this common point is called the output and fed to the oscilloscope. The other end of one gate is connected to ground, while the other end of the second gate is connected to a variable reference voltage, such as would be generated by a constant-voltage source and a precision potentiometer. Now, when the two analog gates in the 4016 are alternately switched on and off, the waveform on the oscilloscope will be a square wave that switches from zero to the reference voltage.

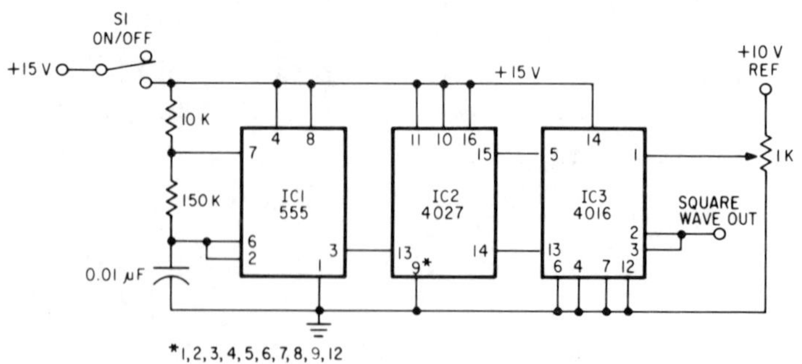

Fig. 5-10.   Oscilloscope calibrator.

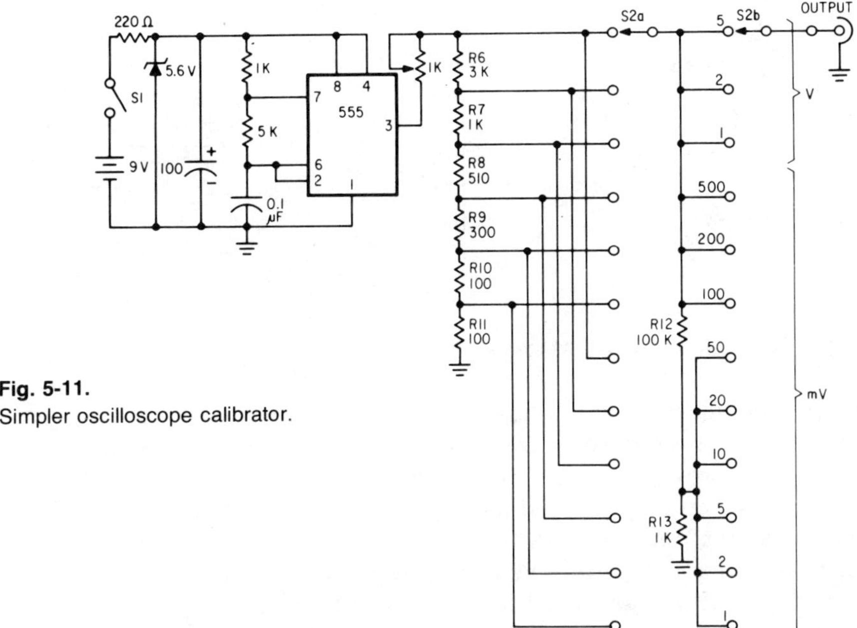

Fig. 5-11.
Simpler oscilloscope calibrator.

5.10   simpler oscilloscope calibrator

A simpler version of the oscilloscope calibrator is seen in Fig. 5-11. This version simply uses a 555 in the astable mode and then feeds the timer output to a two-stage, switched attenuator, to provide variable output amplitudes. Again, the frequency here was chosen to be about 1 kHz.

The output attenuator gives an output voltage range of 5 V to 1 mV. It is constructed from a two-pole, 12-throw rotary switch and eight resistors. One section of the attenuator provides an attenuation of 100:1, while the second provides outputs of 5, 2, 1, 0.5, 0.2, and 0.1 V.

To reduce the possibility of the calibrator giving a false reading, a 5.6-V zener diode is used to supply voltage to the whole circuit. Since the zener is supplied by a 9-V battery, supply voltage fluctuations will be minimal. To further insure the accuracy of the unit, the calibrator should not feed into an impedance of less than 100 kΩ.

To calibrate the calibrator, use the potentiometer to adjust the output voltage to 5 V on the 5-V range.

5.11   triggered sweep adapter*

Still in the area of add-on circuits for the oscilloscope is a handy little device (Fig. 5-12) that adds a triggered sweep capability to any

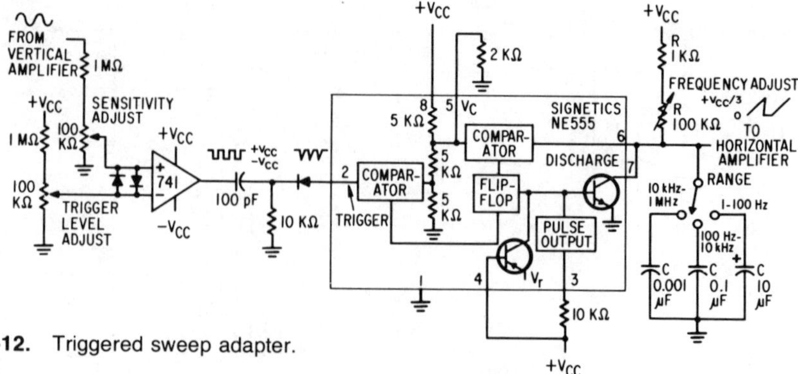

Fig. 5-12.   Triggered sweep adapter.

scope that doesn't already have it. The advantage of adding a triggered sweep to your oscilloscope is that it makes it easier to view logic signals because it provides a rock-steady display.

A 741 op-amp is used as triggering device for this circuit. The noninverting input to the op-amp is connected to the vertical amplifier of the oscilloscope, while the inverting input is connected to a variable reference voltage and used to control the trigger level.

Normally the output of the 741 is at the positive supply voltage level. However, when an input voltage rises above the trigger level as set by the potentiometer, the output of the op-amp goes to the negative supply voltage. This negative transition gets capacitively coupled to the trigger input (pin 2) of the 555, where it causes the output of the timer to go high. At the same time, this allows the timing capacitor to start charging through resistor R.

The timer resets itself when the capacitor voltage reaches $\frac{2}{3} V_{cc}$ and is then ready for the next trigger pulse. Thus, the frequency of the ramp voltage is determined by the RC combination and may vary from 1 Hz to 1 MHz. The ramp voltage produced across the capacitor is then coupled to the horizontal amplifier of the scope.

5.12   frequency meter with overrange indicator*

By making use of a 556 IC, which is composed of two 555 timers in a single package, an overrange indicator can be economically added to an analog frequency meter. A 555 can be used alone as a monostable multivibrator that is triggered by the frequency to be measured. To provide unambiguous measurements, however, the meter described here uses a second timer to flash a warning light whenever the input exceeds the maximum frequency setting. Although the technique of using monostables in analog frequency meters is not new, the use of new circuit developments makes the design economical and easy to implement (Fig. 5-13).

When the range switch on this meter is set to the 50-Hz range, any input frequency from near dc to 50 Hz causes a panel meter to read correctly; e.g., a frequency of 42 Hz produces a meter reading of 42 μA. However, the meter reading is incorrect when the input frequency exceeds 50 Hz, and therefore a light-emitting diode overrange indicator flashes. If the range switch is then moved to a setting higher than the frequency, the LED stops flashing and the meter again indicates correctly. For example, a 300-Hz signal would be measured on the 500-Hz range, and the meter would show 30 μA.

In the meter diagrammed here, the upper portion of the circuit measures the frequency and has the 50-μA panel meter as its readout. The lower portion provides the overrange indication and has the LED as its warning light. These two portions of the circuit are driven by a common input.

The input signal is a rectangular pulse train; the pulses are differentiated to produce the negative spikes that are needed to trigger the timer. For a sine-wave or sawtooth input signal, a Schmitt trigger might be used to generate the negative impulses.

When pin 6 of the frequency-measurement monostable is triggered, pin 5 goes high. It stays high and delivers current for a time equal to 1.1R1C1. This positive output pulse appears once for every

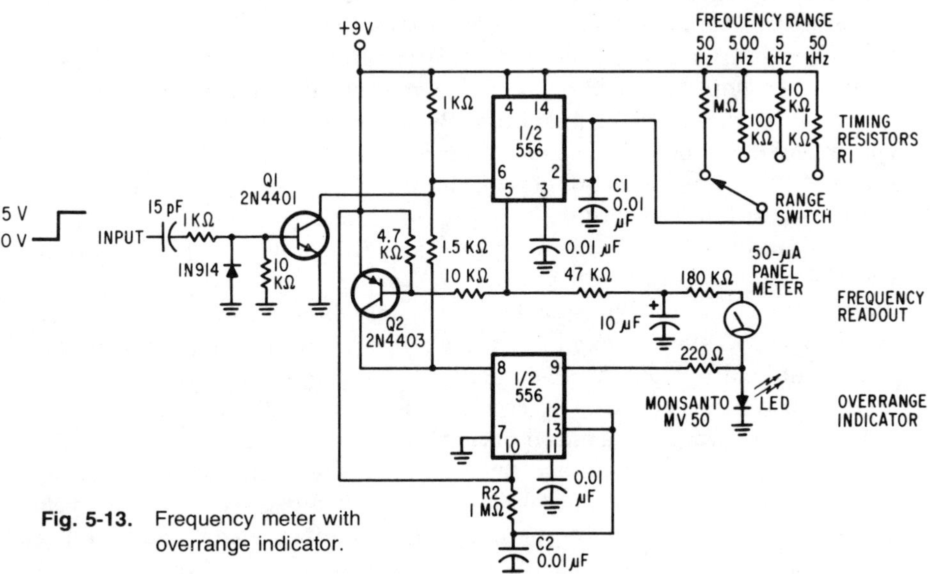

Fig. 5-13. Frequency meter with overrange indicator.

cycle of the input frequency (unless the trigger impulse arrives while the output at pin 5 is already high). The current pulses, smoothed by the 10-μF capacitor, provide an average value that is shown on the microammeter.

At low frequencies, the output pulses are well separated, so the average current is low. At higher frequencies, however, they are closely spaced and approach a duty factor of about 95% at the upper frequency limit set by the range switch. Average current thus increases as the frequency increases. Resistors in the output circuit are chosen so that the average current is 50 μA at the maximum frequency in each range.

If the input frequency exceeds the meter range, a trigger spike arrives while the output is already high. As a result, that input cycle is not counted, so the frequency meter indication is erroneous.

To warn that trigger impulses are arriving while pin 5 is high, pin 5 is also connected to the base of pnp transistor Q2. When pin 5 is low, Q2 conducts and holds pin 8 high, thus preventing the warning-indicator monostable from being triggered. But when pin 5 is high, Q2 is turned off; a negative input spike that reaches pin 8 therefore can trigger an output from pin 9 that flashes the LED. The duration of the flash is 1.1R2C2.

5.13   audio-frequency meter*

The 555 IC timer can easily be made into an inexpensive linear frequency meter covering the audio spectrum. The 555 is used in a monostable multivibrator circuit. The monostable puts out a fixed time-width pulse, which is triggered by the unknown input frequency.

Referring to Fig. 5-14, transistors Q1 and Q2 are used as an input Schmitt trigger. The unknown frequency input is clipped between 9 V and ground by these transistors. Positive feedback is used to insure the waveforms have fast, clean edges. The output of Q2 is a square pulse with the same frequency as the input signal. The output of Q2 is differentiated by C2 and R2 to provide a short pulse for the 555. A small signal diode is connected across the differentiator to insure that the 555 input signal never exceeds 9 V.

Since a Schmitt trigger drives the monostable, square, sine, and ramp type waveforms may be used at the input to the frequency meter. A nominal voltage of 1 V rms is required to trigger the Schmitt circuit.

The range scale timing resistors R3, which determine the mon-ostable pulse width, are small potentiometers mounted directly on the circuit board. These pots are used to calibrate each frequency scale.

The monostable output is a fixed-width pulse with a duty cycle dependent on the input frequency. Thus, by integrating or averaging

* Hinkle, G., "IC Audio Frequency Meter," *73 Magazine*, Holiday 1976 issue, p. 61.

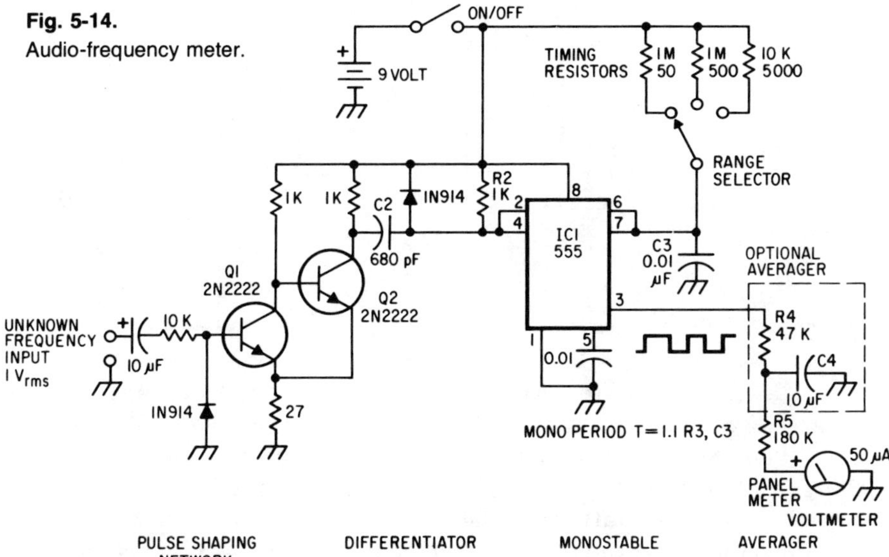

the output waveform, a dc voltage is developed. This voltage is directly related to frequency. Resistor R4 and capacitor C4 are used as pulse averagers, important on the lower range setting. As the input frequency increases, the panel meter itself can act as a waveform averager. Input frequencies greater than 50 Hz will be averaged by the meter fairly well; however, at lower frequencies the meter will respond to each cycle of the unknown frequency input. The meter is used as a high-impedance voltmeter.

The range scales are set up in decades. To calibrate each scale, a standard input frequency is connected to the input. About 1 V rms is needed to trigger the first stage. The monostable output pulse period must be less than the maximum input frequency to be measured on each scale. With the maximum input frequency applied, each range potentiometer should be adjusted until the value of the input frequency corresponds to the full scale meter reading.

Once calibrated, this frequency meter should read within 5% of full scale. The useful frequency range of the meter is from tens of hertz to well over 50 kHz.

5.14   simple capacitance meter*

The circuit in Fig. 5-15 shows how the 555 timer can be used to make a linear-scale capacitance meter. Two characteristics of the 555 make it well-suited for this application. The first is its wide frequency range; the second, its high output current (both sourcing and sinking).

* Horton, R., "555 Timer Makes Simple Capacitance Meter," *EDN*.

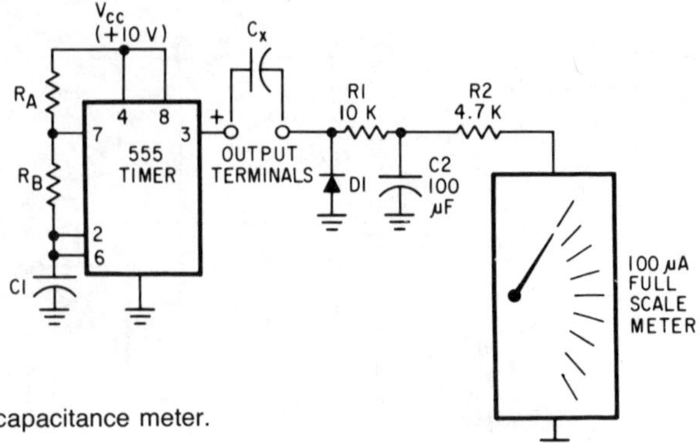

Fig. 5-15.   Simple capacitance meter.

The circuit operates as follows: the timer is connected as an astable square-wave multivibrator, with frequency determined by R_A, R_B, and C1. When the output of the 555 is high, the unknown capacitor, C_X, charges almost to V_{cc} with the current passing through the meter circuit. When the output of the 555 goes low, C_X discharges to approximately 0 V through D1. Since charge on a capacitor is equal to capacitance times voltage, the net charge, Q, per cycle equals $C_X V$. The effective current through the meter circuit (I_{eff}) then equals the charge per cycle times frequency, or I_{eff} equals Qf. Thus, $I_{eff} = C_X Vf$, and the meter scale is linear. For $V_{cc} = 10$ V, Table 5-1 shows full-scale capacitance range versus frequency for a 100-μA meter.

R1, R2, and C2 damp the meter movement enough so that at f=10 Hz and above the meter appears to read a constant, even though the charge is a series of pulses. At f=1 Hz, the pointer bounces slightly but still allows a rough interpolation. The meter damping could be increased if desired, but this would lengthen settling time.

Since the time needed to charge the unknown capacitor through the meter circuit is much longer than the time needed to discharge through D1, an asymmetrical duty cycle is desired. With $R_A \approx R_B$ the charging time is twice the discharge time.

Since the unknown capacitor is charged to 10 V during part of the cycle, it is important that polarized capacitors be oriented properly.

Table 5-1. Timer Frequency Requirements

C_X (Full-Scale)	Frequency
100 pF	100 kHz
1 nF	10 kHz
10 nF	1 kHz
0.1 μF	100 Hz
1 μF	10 Hz
10 μF	1 Hz

Also, because the meter reading is directly proportional to the supply voltage, a regulated supply should be used for best accuracy. If this is done, it is simple to make accuracy better than you can read on a meter.

5.15   digital capacitance meter

In many applications, it is convenient to have a direct digital readout of capacitance. If you own a digital counter and have three 555 IC timers, you can make an adapter unit that will allow your digital counter to be used as a digital capacitance meter.

As can be seen in Fig. 5-16, the first timer (T1) is configured to operate as an astable multivibrator with a frequency of 1 Hz. The output of this timer is fed to the trigger input of the second timer (T2). T2 is connected so that it operates as a monostable multivibrator. The width of the output pulse of this monostable is determined by the time constant $R3C_X$, where C_X is the unknown capacitor. When T2 is triggered, the discharge transistor that normally shorts the timing capacitor C_X is disabled and the capacitor starts to charge through R3. When the voltage across C_X reaches $\frac{2}{3}V_{cc}$, an internal comparator triggers the internal flip-flop in the 555, and the timing capacitor is once again shorted.

During the time that the timing capacitor is charging, the output pin of the timer goes high. When the capacitor is discharged, the output goes low again. It is this output signal from pin 3 of T2 that is used to gate the third timer IC, which operates as an astable multivibrator at 100 kHz. Pin 4, the reset pin of T3, is not tied to the positive supply rail as it generally is in astable applications but is instead connected to the output pin (3) of T2. Thus, T3 is generally not oscillating because the normal condition for the output of T2 is low. However, once a second T2 is triggered by T1, so T3 is gated "on" once a second and generates bursts of 100-kHz signals.

The direct measurement of capacitance is possible because the duration of the burst, and thus the number of pulses in the burst, is directly proportional to the capacitance under test. The relationship for this proportionality is $t = 1.1R3C_X$.

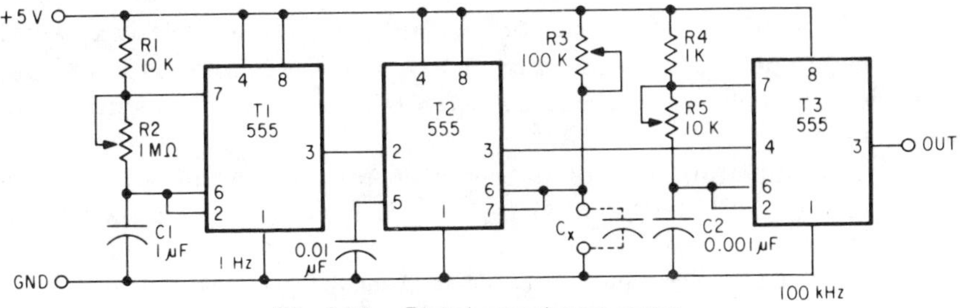

Fig. 5-16.   Digital capacitance meter.

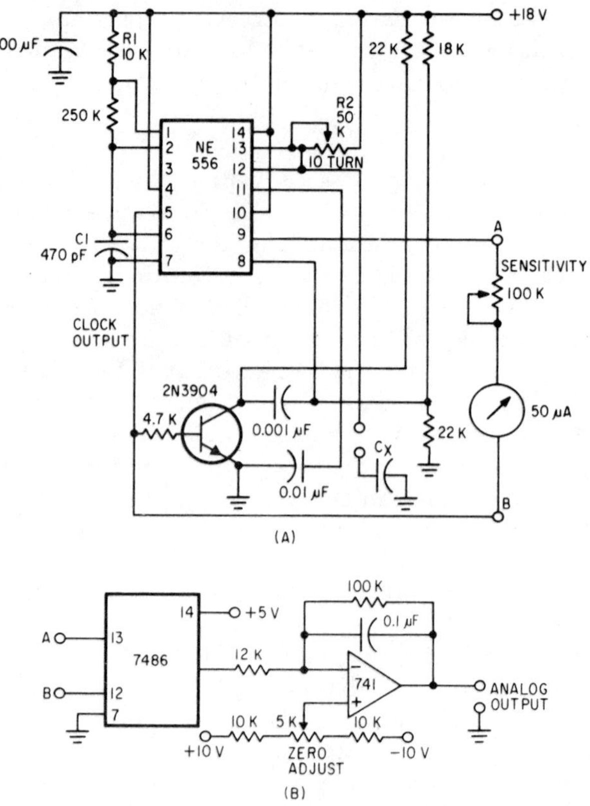

Fig. 5-17. Sensitive capacitance meter.

This circuit will measure capacitance values between 1,000 pF and 10 μF with good accuracy. The unit is calibrated by taking a capacitance of known value and by adjusting R3 so that one output pulse is equal to 100 pF. Also, the two oscillator frequencies should be adjusted with a frequency counter.

5.16   sensitive capacitance meter*

A single 556 timer, or two 555s, can be used to make an extremely sensitive capacitance meter (Fig. 5-17). The first timer is wired as a clock, which operates at about 100 kHz; the second operates in a mono-stable mode.

The monostable circuit is triggered by the leading edges of the clock pulses. A 2N3904 transistor inverter couples the pulses to the

---

* Blair, D. G., "Timer Chip Becomes Meter That Detects Capacitance Changes of 1 Part in 10⁶," *Electronic Design*, March 5, 1976, p. 70.

monostable's input. The monostable's "on" time is determined by the time constant of the 10-turn pot, R2, and the unknown capacitance C_X.

A 50-μA meter measures the difference in pulse length between the clock pulses and the monostable pulses at points A and B in Fig. 5-17A. Alternatively, a proportional dc analog voltage can be obtained by use of an exclusive-OR gate that provides the difference between the two pulse lengths, and an integrator circuit to convert this difference into an analog signal (Fig. 5-17B).

The circuit can detect capacitance changes as small as 1 part in 10^6 at 10 nF, and it can detect changes over a range of 50 pF to 50 nF with the values of R1, R2, and C1 used in Fig. 5-17A. The circuit may be used as a readout for such applications as capacitive pressure transducers and liquid-level gauges, or wherever you may need to monitor small capacitance changes of a large capacitance. If desired, you can make absolute capacitance measurements by calibrating the precision 10-turn pot against known capacitors.

5.17 resistance-capacitance bridge

Very often it is possible to get a good buy on a batch of unmarked components. With the aid of a 555 astable and a bridge circuit, the values of unmarked resistors and capacitors can be easily identified. This bridge has six ranges that cover capacitors from about 10 pF to 10 μF and resistors from about 70 Ω to 10 MΩ (Fig. 5-18).

The theory of operation of this circuit is fairly simple. The 555 produces an audible signal whose frequency can be varied to suit the listener by adjusting potentiometer R3R4. The output of the astable is coupled to the bridge circuit via C3. To explain the operation of the bridge, it will be assumed that S1 is connected to one of the internal reference resistors, but it could just as easily be connected to a capacitor and produce the same results.

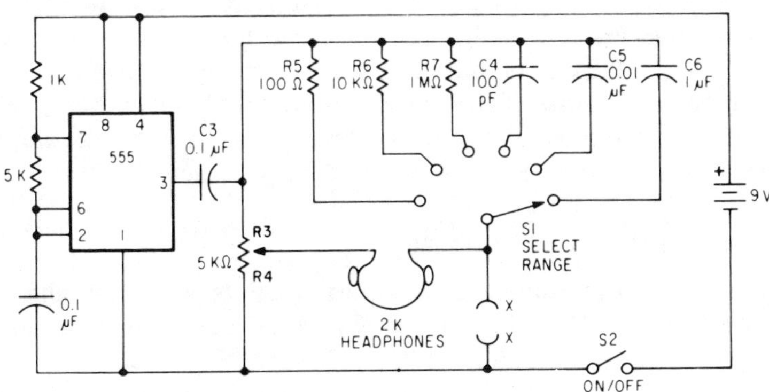

Fig. 5-18. Resistance-capacitance bridge.

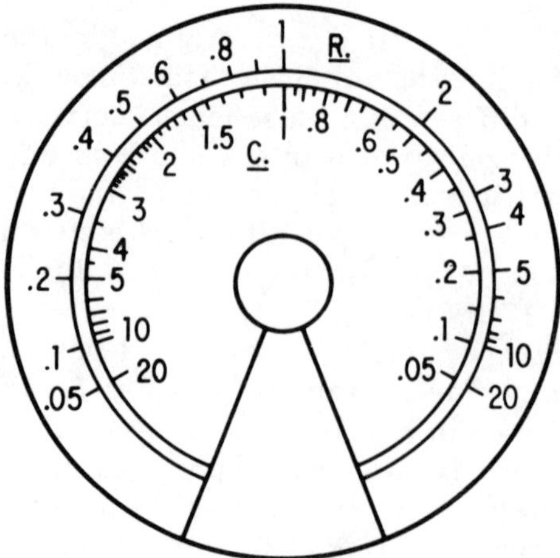

Fig. 5-19. Calibrated scale for resistance-capacitance bridge.

With the switch connected to R6, an unknown resistor R_X must be connected to the test terminals x-x. If an unknown capacitor is to be checked, S1 connects to one of the reference capacitors.

Now, when the ratio R3/R4 = R6/R_X, the bridge is said to be balanced and thus no current will flow through the headphones, and no audio is heard. For any other value of the components, the bridge is unbalanced and a tone is heard in the earphones because a current flows through them. When checking an unknown component, S1 is used to select a range and the type of component, while pot R3R4 is used to balance the bridge and indicate the component's value. This is done by adjusting R3R4 until no sound is heard. If electrolytic capacitors are being checked, it may not be possible to eliminate the tone entirely because of the capacitors' leakage.

The accuracy attainable by this circuit is a function of the accuracy of the reference resistors and capacitors used. R5, R6, and R7, as well as C4, C5, and C6, should all be 1% components for best accuracy, but devices with lower tolerances can be used. After you build the unit, it will not be necessary to calibrate the scale for R3R4, since the scale printed in Fig. 5-19 can be photocopied, or traced and used. Just make sure that the endpoints of the potentiometer rotation coincide with the endpoints on the scale. Another important point: the potentiometer used for R3R4 must be a linear one and not a log or semilog unit.

5.18   two-phase clock generator*

In many applications, it is necessary to generate a two-phase clock signal with pulses that do not overlap. This can be done with a 555 timer, a 7473 flip-flop, and a 7402 NOR gate (Fig. 5-20).

* Schlitt, G., "Monolithic Timer Generates 2-Phase Clock Pulses," *EDN*.

The 555 is configured as a conventional astable and its output is fed to a NOR gate that is connected as an inverter. This signal is then used to drive the toggling input of a flip-flop as well as one input of each of the two remaining NOR gates. Each of the two complementary outputs of the flip-flop goes to one of the two output NOR gates and they are used to control the phases that are permitted to pass through the NOR gates.

The pulse width of the output pulses is easily adjusted by varying the values of R1 and R2 as in a conventional astable circuit.

5.19   multiphase clock generator

While the previous circuit can only generate a two-phase non-overlapping clock signal, this circuit can generate a seven-phase non-overlapping clock signal (Fig. 5-21). The number of phases available is programmable and can be selected by simply rotating S1 to the desired position.

The heart of this multiphase clock is an MSI (for medium scale integration) IC called the 74155. This is a three-to-eight lone multiplexer. The multiplexer is driven by a 7490 decade counter and the number of phases produced is determined by where the connection for resetting the 7490 to zero is placed. If it is placed on line 8 as in Fig. 5-21, then seven phases will be produced. If it is placed on line 2, then only one phase will be produced.

The decade counter is driven by a conventional 555 astable circuit. An important point to remember, however, is that the output frequency of the generator will be equal to the output frequency of the astable divided by N + 1, where N is the number of output phases.

5.20   variable audio-frequency generator

You can build a simple square-wave audio-frequency generator for checking out audio equipment by combining a 555 astable with a

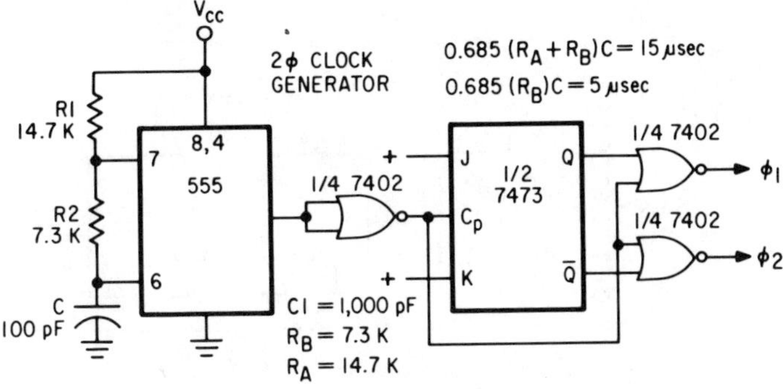

Fig. 5-20.   Two-phase clock generator.

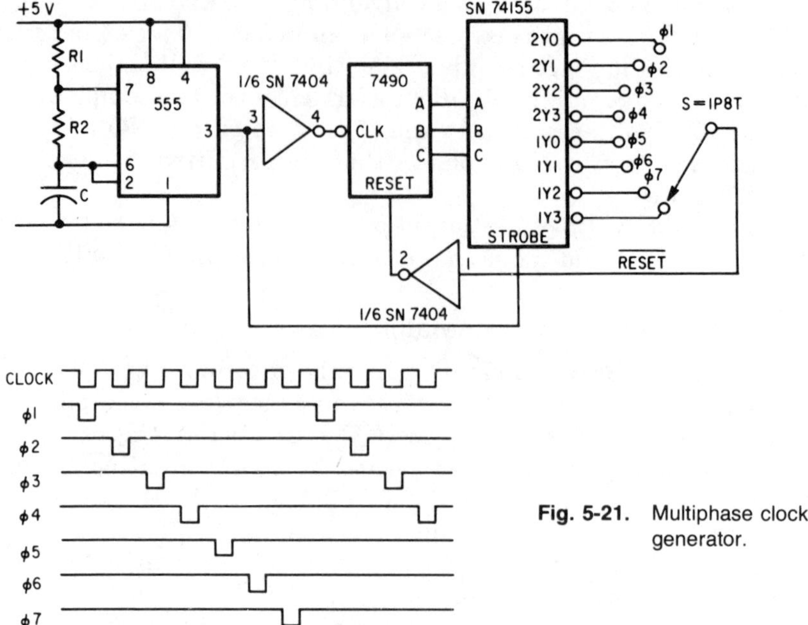

Fig. 5-21. Multiphase clock generator.

flip-flop and some gates (Fig. 5-22). The 555 is hooked up to work as an oscillator with a frequency range of between 18 Hz and 22.5 kHz, a range slightly larger than that which is audible to the average person. The output of the oscillator is then fed to two places. It goes directly to a buffer and from there to the output, and it also goes to a dual flip-flop, where it gets divided first by 2 and then by another 2. From the flip-flop, the two divided signals go to buffers made from logic gates and then to a switch where the desired frequency can be selected. The switch thus makes it possible to select f, f/2, or f/4, where f is the frequency indicated on the dial as determined by potentiometer R3. A frequency counter should be used to calibrate the dial.

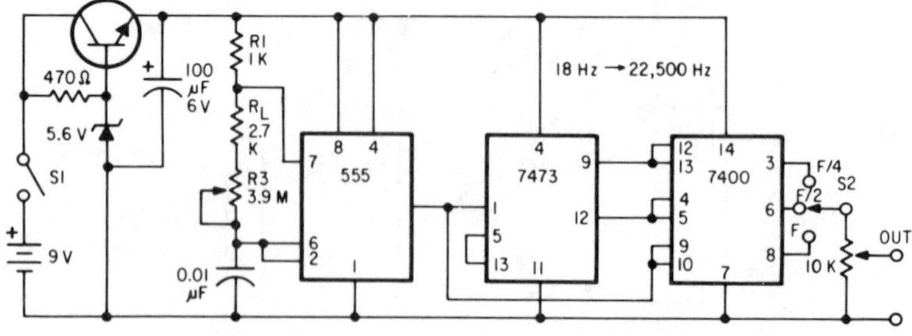

Fig. 5-22. Variable-audio-frequency generator.

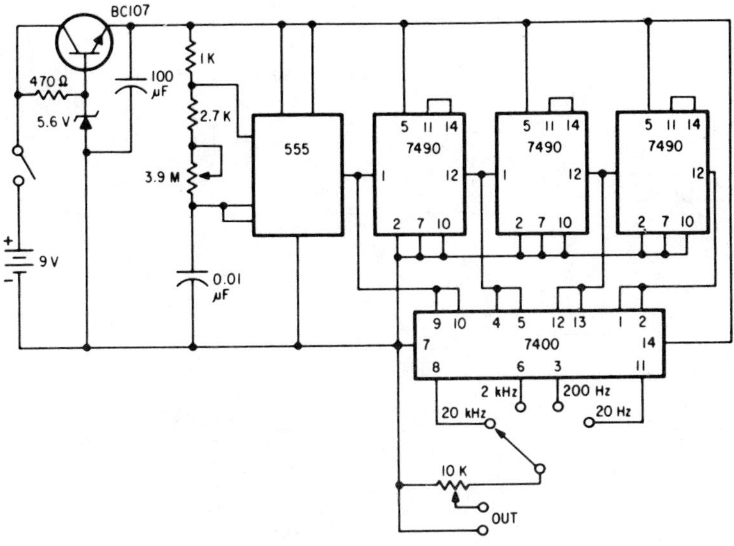

Fig. 5-23. Improved audio-frequency generator.

Power is supplied to the generator by a 9-V battery whose voltage is reduced by a regulator to the 5 V required by the TTL logic circuits.

5.21 improved audio generator

While inexpensive and easy to build, the audio generator in the previous section has some disadvantages. The key one is that, because one potentiometer is used to cover the entire audio range, it may be a little difficult to get a specific frequency with a high degree of accuracy.

This problem can be remedied quite easily by simply feeding the output of the astable to a chain of divide-by-ten circuits and tapping off the signal after each division as in Fig. 5-23. This arrangement will give four ranges that have the following maximum frequencies: 20 kHz, 2 kHz, 200 Hz, and 20 Hz. As with the previous circuit, the outputs are then fed to buffer gates and then to a selector switch. From there it goes to a potentiometer that adjusts the amplitude from 0 to 5 V.

5.22 programmable pulse generator*

An inexpensive pulse generator programmable over a wide frequency range (10,000:1) can be constructed using a 555-type timer. The circuit has good linearity (3% at room temperature without trimming),

---

* Laughlin, E. G., "Inexpensive Pulse Generator is Logic Programmable," *EDN*, Aug. 20, 1974, p. 92.

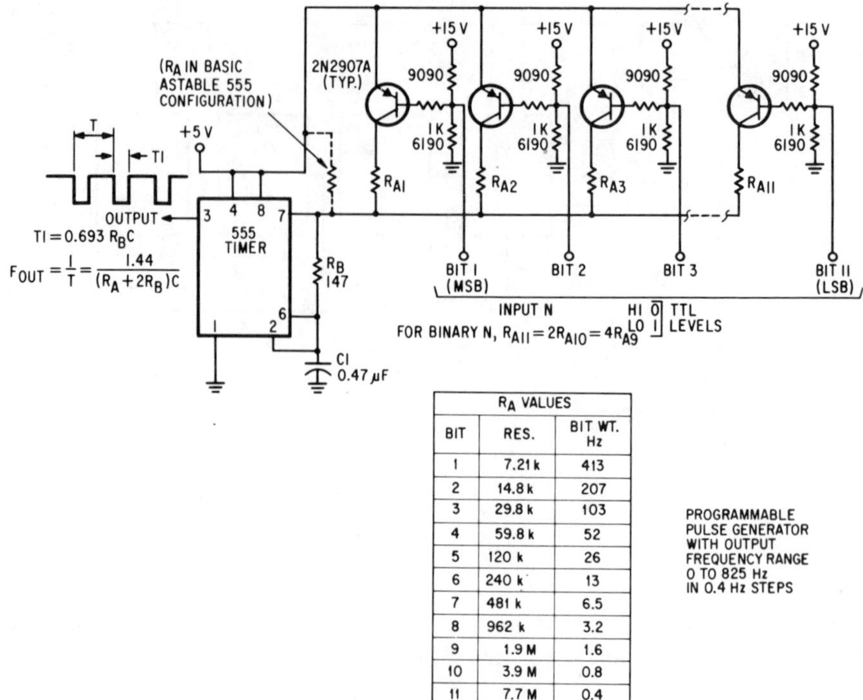

$$TI = 0.693 R_B C$$

$$F_{OUT} = \frac{1}{T} = \frac{1.44}{(R_A + 2R_B)C}$$

R_A VALUES		
BIT	RES.	BIT WT. Hz
1	7.21 k	413
2	14.8 k	207
3	29.8 k	103
4	59.8 k	52
5	120 k	26
6	240 k	13
7	481 k	6.5
8	962 k	3.2
9	1.9 M	1.6
10	3.9 M	0.8
11	7.7 M	0.4

Fig. 5-24. Programmable pulse generator.

and can be designed to operate over the entire frequency range of the 555 (to about 200 kHz).

This circuit (see Fig. 5-24) is basically a 555 connected in its astable mode with timing resistor R_A replaced by a group of timing resistors and switching transistors. The circuit shown operates from 0 to 825 Hz and has TTL-compatible inputs and output. However, simple changes can give higher frequencies and compatibility with a wide range of inputs and outputs.

When any bit input is high (0), its associated transistor is turned off. When that input is low (1), the transistor is on and operating in its saturated mode. This allows C1 to charge through the associated R_A and R_B. If more than one transistor is on, the charging path is through the parallel combination of their associated R_As and R_B. The effective value of R_A then becomes R_A (LSB)/N, where N is the binary input. If $2R_B$ is selected small enough to be negligible when compared to all R_As, the output frequency is directly proportional to N. In any case, output pulse width (T1) is maintained constant over the frequency range, since it is dependent only on R_B and C.

Using the biasing technique shown, an input of $N=0$ turns all transistors off hard enough to produce an output frequency which is essentially zero. In fact, in this condition a period of several minutes is required for C to charge sufficiently to trigger the 555. Output frequency change is "smooth" throughout the operating range. Spurious pulses are not generated when N changes, and the circuit can transition from one value of N to any other value of N between two output pulses, or during one output pulse.

5.23   programmable pulse-width generator

In applications in which it is necessary to have a low-frequency (less than 2 kHz) square waveform with a programmable output pulse width, it is possible to use the following circuit (Fig. 5-25) to produce a waveform whose width can be programmed via BCD switches or logic signals.

As in most of the generator circuits, we start out with a basic astable multivibrator and add some things to it. In this case what we add is a CMOS divide-by-100 dual BCD counter (4518) and two BCD comparators (4063).

The reason that this circuit is limited to frequencies of about 2 kHz and less is that the output frequency is equal to the input frequency divided by 100. And the output pulse width can be varied from 1% to 100%. The BCD information is applied to the A1, A2, A4, A8, A10, A20, A40, and A80 inputs on the two comparators.

5.24   staircase generator

A simple method of generating a staircase waveform is seen in Fig. 5-26. Here, the output of a 555 oscillator is fed to an eight-bit

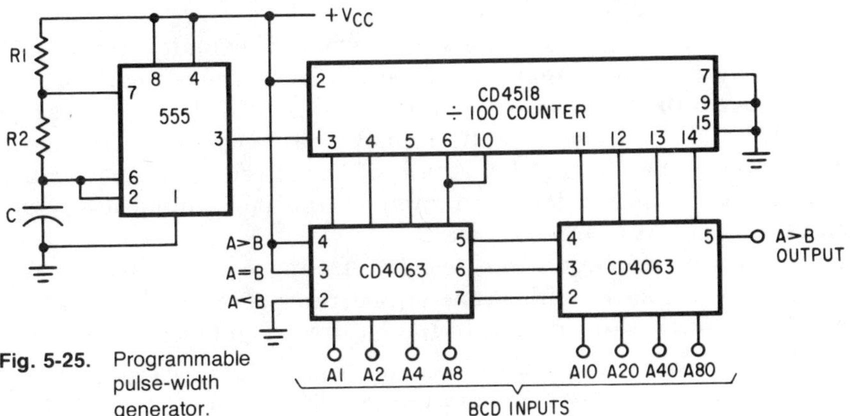

Fig. 5-25.   Programmable pulse-width generator.

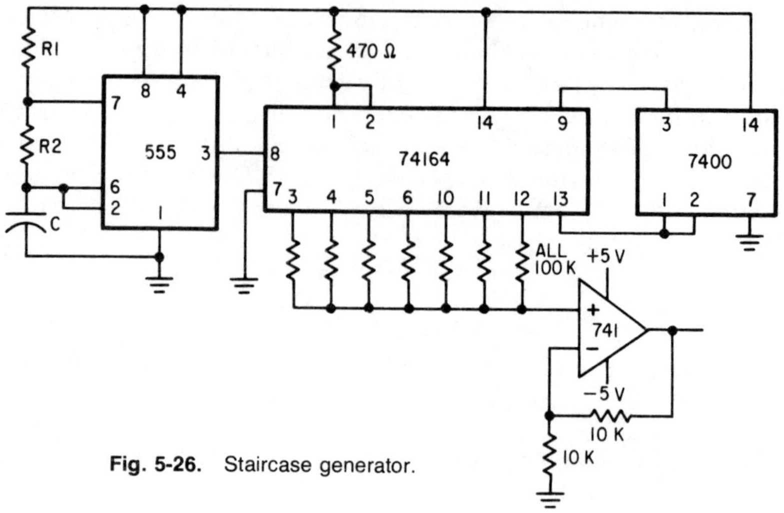

Fig. 5-26.   Staircase generator.

parallel output shift register (74164). Each output of the shift register is in turn connected to a 100-kΩ summing resistor, and the common end of the resistors is fed to an operational amplifier.

By taking the signal provided at the eighth bit, inverting it, and applying it to the clear input of the shift register, the waveform can be repeated continuously. To produce a step waveform with fewer steps, simply connect the gate input to a lower-order bit. For example, if a four-step waveform is desired, the gate input gets connected to the fifth bit.

If a waveform with more steps is needed, two 74164s can be used. In this case, pin 13 of the first 74164 gets connected to pin 8 (clock input) of the second 74164.

5.25   power ramp generator*

A 555 timer buffered by a three-terminal regulator, the LM317, can generate a linear ramp to a 1-A output level. The normal 555 pulse output is available at pin 3 of the timer (Fig. 5-27).

The LM317 is used simultaneously as a precision current source and power buffer. As a buffer, the LM317 delivers an output voltage that is 1.25 V above the linear ramp of the R_tC_t timing node. As a current source, it charges C_t, via R_t.

Since in this circuit, the resistive timing network usually used with a 555 is replaced with a constant-current source, the timer's pulse width is directly proportional to the V_{cc} line. The pulse width can be

* Jung, W. G., "A Power Ramp Generator Delivers an Easily Adjustable 1-A Output," *Electronic Design*, March 1, 1976, p. 66.

linearly modulated, if desired, by adjustment of the voltage at the 555's control terminal, pin 5, shown dotted. Without this adjustment pin 5 is referred to $\frac{2}{3}$ of V_{cc}, so with a 15-V supply, the pulse-width equation reduces to $T \simeq 8\ R_t C_t$.

The 555 also can be driven at pin 5 from a low-impedance source. Such operation would be independent of V_{cc} and useful for pulse-width modulators, PLLs, and similar systems.

Resistor R_t is best kept in a range of 100 Ω to 10 kΩ, while C_t should be as small as practical for the desired pulse width.

5.26   pulse pattern generator

A handy thing to have around when you're testing digital circuitry is a means of applying different patterns of pulses to the item under test. To do that, a four-bit binary counter (7493) and a 1-out-of-16 data selector (74150) are used in conjunction with an astable-configured 555. This circuit (Fig. 5-28), will make it possible to generate a 16-bit programmable pattern.

The 7493 is used as a ripple-through counter and its outputs are used to address the data-selector chip. So that for each different pattern (there are 16 of them) of bits at the counter's output, a different line of the selector is addressed. The 16 input lines to the multiplexer are connected to single-pole, single-throw switches that are used to program the bit pattern. The switches are connected to ground, and closing a switch is the equivalent of applying a low logic signal to the input.

In operation, the multiplexer scans each of the inputs in sequence and outputs whatever it finds there in an inverted form.

5.27   simple sine-wave synthesizer

Although the 555 is only capable of generating pulses, ramps, square waves, and triangular waves, it is possible to use the timer with

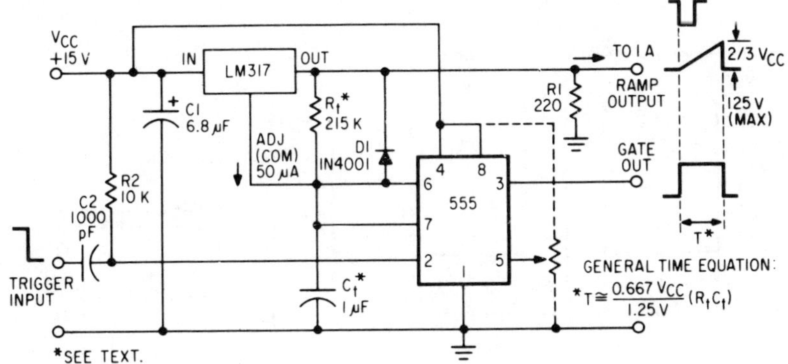

Fig. 5-27.   Power-ramp generator.

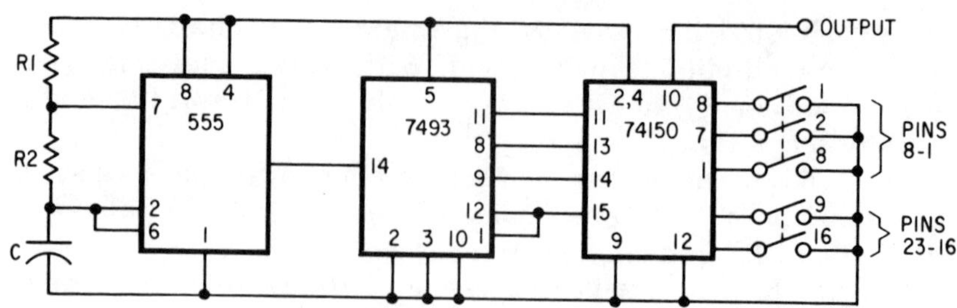

Fig. 5-28.   Pulse-pattern generator.

additional circuitry to generate sine waves or close approximations to them.

. A close approximation of a sine wave can be produced if two signals, one of frequency f and the other of frequency 3f, are added 180° out of phase with each other with a weighting of ⅓ and 1. By doing this, the third harmonic of the f square wave is canceled, yielding a good approximation of a sine wave.

This is done by using the circuit in Fig. 5-29. The astable is used to generate a frequency that is three times the value of the desired sine-wave frequency. The output of this circuit can be improved if desired, by using a simple low-pass filter that cuts off at frequency f.

5.28   better sine-wave synthesizer

A closer approximation to a sine wave can be achieved by increasing the number of steps used to make the approximation. In Fig. 5-30, a CMOS presettable divide-by-N counter (4018) is used to form a 10-step sine-wave synthesizer

A major difference between this generator and the previous one is that this one requires that the 555 oscillator be set to operate at a frequency that is 10 times the desired sine-wave frequency.

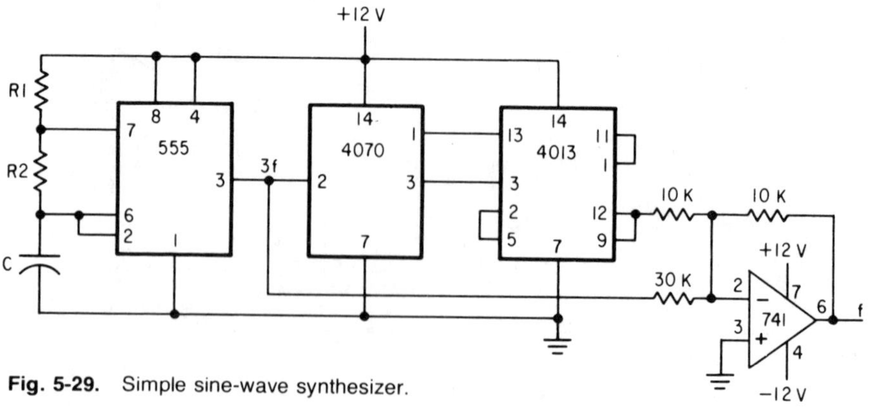

Fig. 5-29.   Simple sine-wave synthesizer.

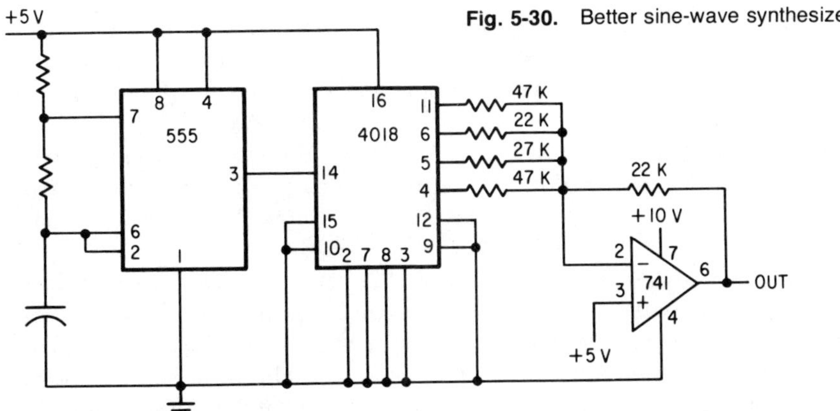

Fig. 5-30. Better sine-wave synthesizer.

As with circuit 5.27, this one uses an operational amplifier to sum all of the outputs together.

5.29 all-purpose tester

Here is a test instrument that includes almost everything but the kitchen sink (Fig. 5-31). It has the capability of measuring voltage, current, and resistance. It also features an audible continuity check, signal injector, and a warning indicator that tells you when the instrument is set on the ohms or milliamps scale. The reason for this is that many a good meter has burned out because someone tried to measure a voltage before checking to see that the meter wasn't on one of these very sensitive scales.

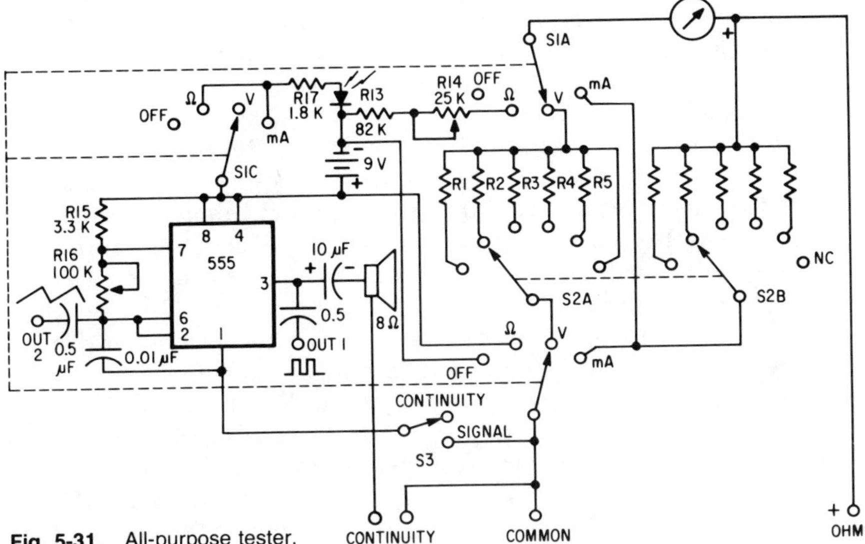

Fig. 5-31. All-purpose tester.

Since most electronics works on low voltages, the voltage ranges have been selected as follows: 50, 10, 5, 1, 0.5, and 0.1 V. Similarly, a number of useful low-current ranges are provided. These are: 50, 10, 1, 0.5, and 0.1 mA.

The instrument is a little less flexible when it comes to measuring resistances. The reason for this is that, since one must measure from zero to infinity, the scale markings get very crowded toward the high end.

The next two features of this unit are almost never found on commercially available units. They are the audio signal injector and the continuity tester. The signal injector makes available two types of waveforms for injection. A pulse waveform is the first and the second is a sawtooth waveform. This makes troubleshooting audio equipment very easy. The waveforms are both produced by the 555 portion of the instrument, and their frequency is adjustable throughout most of the audio range.

The 555 is also used for continuity testing. In this case, a speaker is connected to the output of the timer.

The values for the voltage and current ranges are given for a 100-μA meter with an internal resistance of about 1,050 Ω. Not all 100-μA meters have this resistance (don't try to measure it with an ohmmeter or you'll burn out the microammeter). The internal resistance of the meter you purchase should be listed on it or in the literature that comes with it. If the resistance of your meter is different, new values for R1 through R6 will have to be calculated according to the following formula: $R = (V/I) - r$, where R is the particular range resistor, V is the maximum voltage to be measured on that range, I is the full-scale meter deflection current (100 μA in our case), and r is the internal resistance of the meter. Calculated values can be rounded off to the nearest standard value.

To calculate the values of the shunt resistors for the current-measuring portion of the circuit requires the following formula: $R = r/(N - 1)$, where N is the ratio of the maximum range current to the full-scale deflection current. Again, values should be rounded off to the nearest standard values. This rounding off will introduce some errors, but they will be small.

The variable resistor R14 is used to zero the ohmmeter. This is done by connecting the "ohms" and "common" leads together and adjusting the meter for a full-scale deflection. A full-scale deflection is zero for the ohms scale. Calibrating the scale can be done with some known resistors. A 100-kΩ resistor should give a deflection of a little less than half scale (90 kΩ is half scale). Thirty kilohms should give 75% of full-scale deflection, while 10 kΩ should give 90% of full-scale

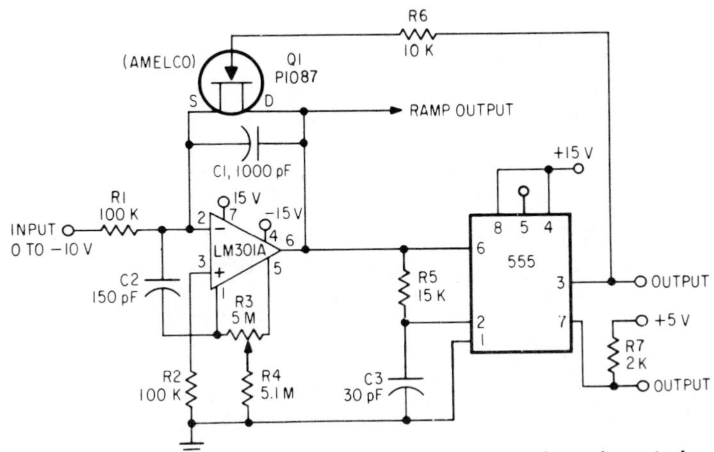

Fig. 5-32.   Negative voltage-to-frequency converter.

deflection. A 300-kΩ resistor should give a little less than 25% of the full-scale deflection, 800 kΩ only 10% of full-scale deflection, and 1 MΩ almost no deflection at all.

5.30   voltage-to-frequency converter*

A simple and accurate voltage-to-frequency (V-to-F) converter can be built with a 555 timer, an LM 301A op-amp, and a few additional parts. Applications for a V-to-F circuit are numerous: in analog-to-digital conversion, analog-data transmission, and voltage-controlled ramp generators, to list a few.

The op-amp, connected as an integrator, generates a positive-going ramp from a negative input voltage (Fig. 5-32). When the ramp reaches two-thirds of V_{cc}, the timer triggers, bringing the output voltage at pins 3 and 7 close to zero. Transistor Q1 then turns on, rapidly discharging C1. Since the discharge time is constant, the linearity of the converter is limited at high frequencies and is controlled by the R5C3 time constant. Resistor R5 and capacitor C3 slow the retrace slope at the input (pin 2) of the 555. When the voltage at this input reaches one-third of V_{cc}, the timer resets, the outputs at pins 3 and 7 return to high levels, Q1 is turned off, and the next cycle begins.

The retrace time for the V-to-F converter is about 1 μsec. This results in less than a 0.2% nonlinearity deviation from the best straight line over the frequency range of 0–10 kHz. At frequencies for which the retrace time can be neglected, the conversion constant is given by $f = 3[V_{in}/2(V_{cc}R1C1)]$. In the circuit shown, $f = (10^3)(V_{in})$.

* Klement, C., "Voltage-to-Frequency Converter Constructed with Few Components Is Accurate to 0.2%," *Electronic Design*, June 21, 1973, p. 124.

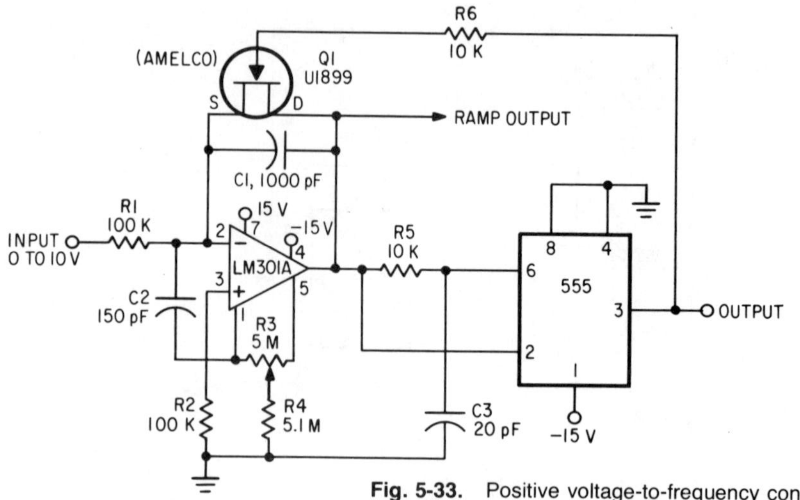

Fig. 5-33.   Positive voltage-to-frequency converter.

If the V_{cc} supply is not sufficiently regulated, the timer's reference voltage at pin 5 can be stabilized if a zener diode is connected to ground and a resistor to V_{cc} and if the resistor is adjusted for the nominal zener current. The voltage at pin 5 then stays at the zener voltage. Capacitor C1 should be a polycarbonate or polystyrene type for best thermal stability. The zero adjust is set by R3 so that the output frequency with a shorted input is about 0.1 Hz. Worst-case shift of this adjustment with a 20°-C temperature change is 1.2 mV, corresponding to a 1.2-Hz frequency shift.

The output at pin 7 of the 555 is directly compatible with DTL and TTL logic circuits. If it's necessary to design for a positive input, the modified version of the circuit shown in Fig. 5-33 can be used. The only drawback here is that the circuit has no direct logic compatibility, unless the −15-V level is used as the common for the logic supply.

5.31   audible logic probe

When working with logic circuits, the need to check the logic level present at a particular location is bound to come up. With this logic probe, you can do that easily, and you don't even have to look at the probe to check the results. All you have to do is listen.

The circuit shown in Fig. 5-34 shows a 555 astable that has some sensing circuitry on its input to differentiate between the two logic levels. The astable is the conventional configuration except the timing capacitor has been replaced by two capacitor-transistor configurations.

In operation, one of the two capacitors (C1 or C2) is switched into the astable circuit and allows the timer to oscillate. C1 is twice C2 so that the difference in the tone is about an octave and clearly noticeable.

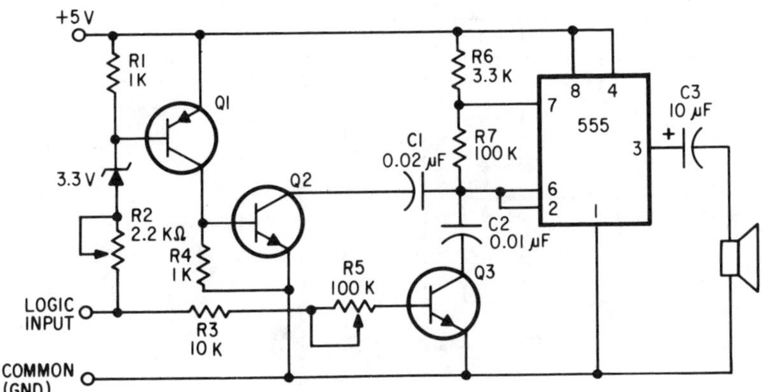

Fig. 5-34. Audible logic probe.

When a logic zero (0.8 V or less) is applied to the probe's input, transistor Q1 will cause Q2 to go into saturation (turn it on). This in turn allows one side of C1 to be effectively grounded and turns on the 555 oscillator, producing a low-frequency tone.

If the input to the probe is a logic high (2.4 V or more) Q3 is turned on and switches capacitor C2 into the oscillator circuit. This produces a higher-frequency tone than that produced for the low-level input. The 0.8-V and 2.4-V threshold levels are set by potentiometers R2 and R5, respectively.

5.32   visual logic probe*

An inexpensive probe for detecting and indicating logic states can be built with the versatile 555 timer. The probe takes full advantage of the timer's high input impedance and its stable and predictable threshold levels. Its power output is capable of driving indicator lamps directly. The probe automatically becomes compatible with TTL, HTL, or CMOS logic by selecting the proper supply voltage (Fig. 5-35).

Input logic signals feed the trigger input (pin 2) of the 555 via two diodes that keep the signals from driving pin 2 all the way to V_{cc} or ground. This drive limiting keeps the 555 from erratic behavior, which is particularly likely for inputs near ground potential, where negative-going transients as small as 50 nsec and reaching only -0.3 V can cause the timer to switch states.

The 555 is used as a comparator, with a threshold of $\frac{1}{3}V_{cc}$ set by the device's internal bias. The output (pin 3) assumes an inverted state relative to the input, and can source or sink up to 100 mA. LED-1 lights continuously on a steady high input; LED-2 on a steady low.

* Klinger, A. R., "Logic Probe Built from IC Timer Is Compatible with TTL, HTL and CMOS," *Electronic Design*, June 7, 1976.

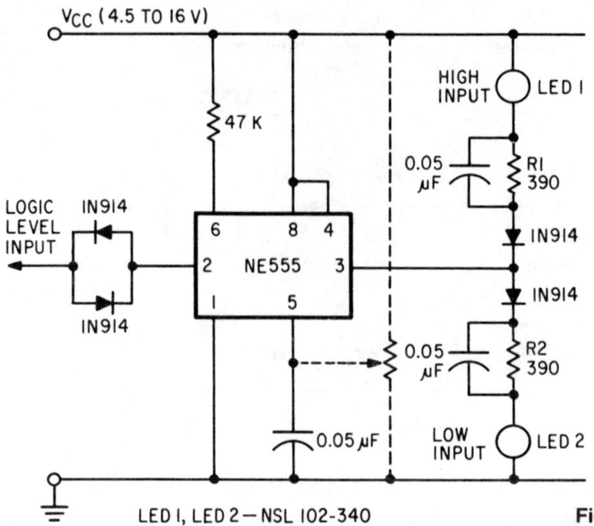

Fig. 5-35. Visual logic probe.

Capacitors across the series-limiting resistors pass current pulses to the LEDs during signal transitions. These pulses momentarily flash the LEDs to show the presence of short input spikes that would otherwise be undetectable. Series diodes protect the LEDs from excessive inverse voltages when the capacitors discharge.

Resistor values chosen for R1 and R2 are a compromise to allow use of widely available 1.6-V, 40-mA LEDs over the full range of the commonly used supply voltages—4.5–16 V. For a single supply level, a compromise isn't necessary. With TTL levels, these resistors could be reduced to 120 Ω; with CMOS or HTL, they can be raised to 820 Ω.

For the time constants shown in the circuit, 300-nsec pulses, spaced 500-μsec apart, remain detectable even at V_{cc} of 5 V. Square-wave inputs light both LEDs equally, which are easily visible to over 500 kHz. The ratio of brightness is an indication of the duty cycle.

Accurate level detection can be added to the circuit, if the reference terminal (pin 5) is connected to a potentiometer wired between V_{cc} and ground. In this way, the threshold of the input (pin 2) can be adjusted from about 1 V to $\frac{1}{2}V_{cc}$. Also, hysteresis can be added by omitting the pin-5 bypass capacitor and providing positive feedback with a resistor from the output (pin 3) to pin 5.

Supply voltage V_{cc} can be borrowed from the circuit under test; thus, the probe easily fits into a penlight case.

5.33   continuity tester

Although a continuity tester was included in circuit 5.29, it may not be clear how to isolate just that portion of the circuit for use by

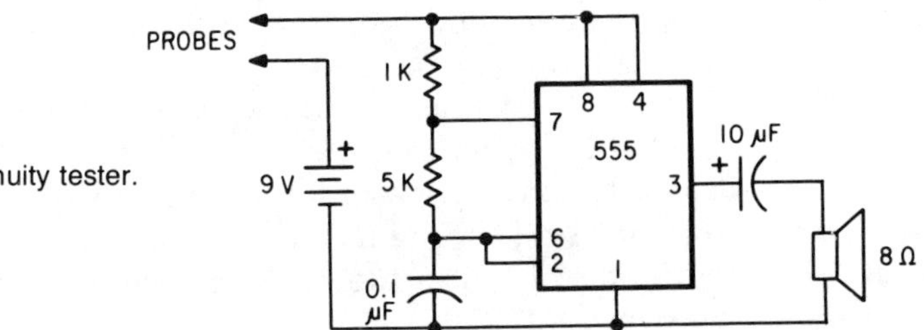

Fig. 5-36.   Continuity tester.

itself. So in Fig. 5-36 is a diagram of the 555 designed to operate as a continuity tester.

Basically, it is an astable multivibrator that has a break in one of the power leads. The two sides of this break are connected to two test probes. When the probes are shorted together, as would be the case in the presence of continuity, the timer oscillates and produces a tone in a loudspeaker. The whole unit can easily be built into a small tube, with a metal tip on one end acting as one probe and a wire with a clip coming out of the opposite end as the other probe.

5.34   sine-wave generator

In circuits 5.27 and 5.28, we saw how it was possible to use a 555 to produce a good approximation of a sine wave, by adding square waves of different frequencies and amplitudes. In Fig. 5-37, we see how to produce a sine wave directly.

In this circuit we take advantage of the fact that, theoretically, a square wave is actually a summation of harmonically related sine waves. Here we do the opposite: we take this summation of harmon-

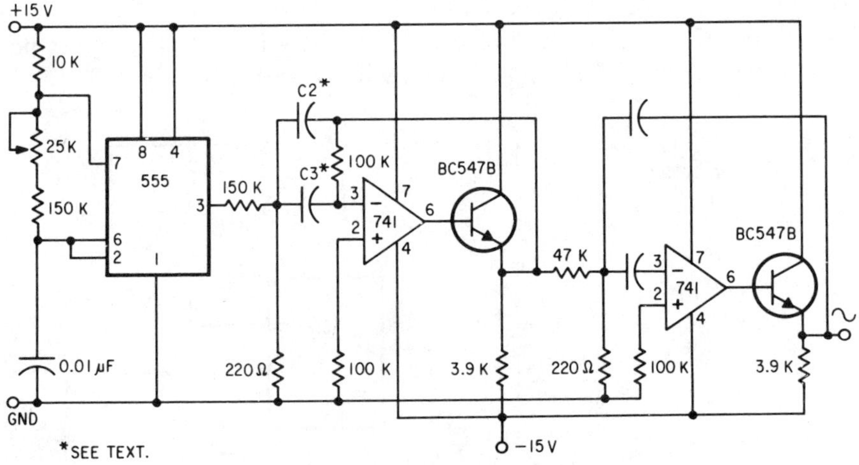

Fig. 5-37.   Sine-wave generator.

ically related sine waves and filter out all of the harmonics, leaving only the fundamental.

The 555 is set to oscillate at between 46 and 68 Hz, depending on the setting of R2. The output of this circuit is then fed into a two-stage, low-pass filter that is designed to prevent frequencies higher than 60 Hz from passing through. Since the 555 is operating at about a 50% duty cycle, distortion in the final sine wave should be minimal and should approach 0.2%.

The frequency of the square-wave generator can be changed according to the standard astable multivibrator formula, and the cutoff frequency of the low-pass filter can be controlled by changing capacitors C2, C3, C4, and C5, which all have the same value, C. This value C can be determined by the following formula: $C(\mu F) = 0.34/f(Hz)$.

5.35 capacitance comparator/leakage tester

A quick and easy way to check the value of electrolytic capacitors is to make use of the fact that, in a monostable circuit, the output pulse width is directly proportional to the value of the timing capacitor.

In the circuit in Fig. 5-38, the 555 is connected to operate as a manually triggered monostable multivibrator. When switch S2 is momentarily closed, it triggers the monostable, whose output lights up the light-emitting diode for a period determined by the timing capacitor (which is unknown) and the series resistor. Since the tester can be used to check large-value capacitors, a range switch has been included. This switch reduces the output pulse width by factors of 10 and 100, so that you won't have to wait several hundred seconds when testing large electrolytics. Just remember to multiply your time reading by the scale factor.

When the LED lights up, simply see how long it stays on. Once you've measured that period of time, simply multiply it by 0.91 to get

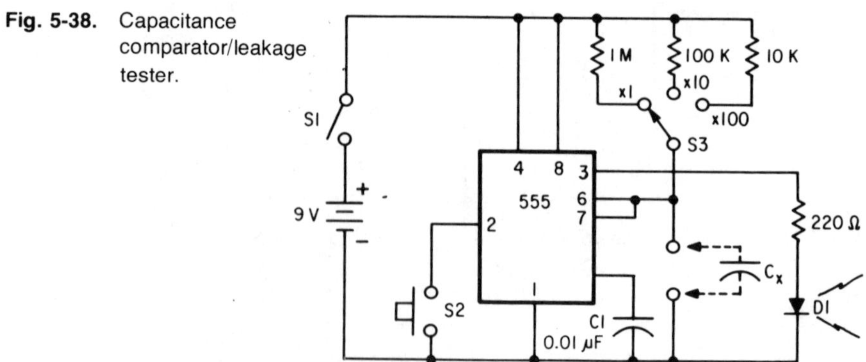

Fig. 5-38. Capacitance comparator/leakage tester.

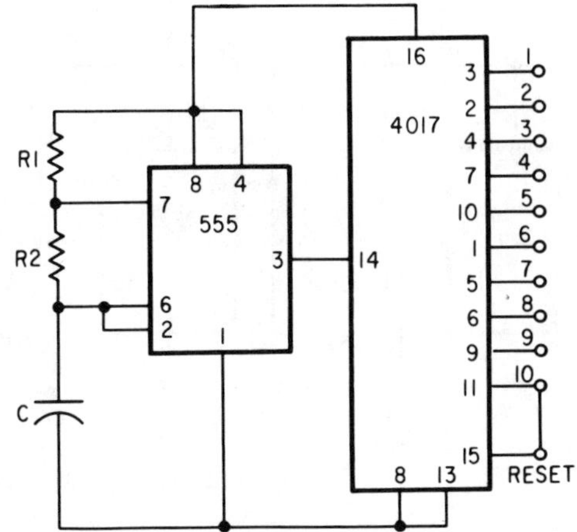

Fig. 5-39.   Test sequencer.

the value of the capacitor in microfarads (remember time must be in seconds). If you are using a range other than $\times 1$, you'll have to multiply the final answer by the scale factor of the range. So, if your unknown capacitor took 10 sec to charge up on the $\times 100$ range, its value would be $0.91(10)(100) = 1,000 \ \mu F$.

If you are checking inexpensive electrolytics, you may come up with some inaccurate readings, this is due to the leakage of the capacitor. But this feature can come in handy. It permits you to take known-value capacitors and check them out. If the measured value is close to the value printed on the unit, then the leakage is very low.

5.36   test sequencer

For cyclical testing of many circuits or devices, it is often desirable to have a way of automatically turning a whole string of devices on and off in sequence automatically. Figure 5-39 shows how to do this. A 555 is once again used in the astable mode to generate a string of pulses. These are fed to the input of a CMOS divide-by-ten counter/decoder (4017). This counter/decoder is really the heart of the sequencer. For each pulse it receives at its input, it generates each of the 10 outputs in turn, for one clock period. By connecting the different outputs to a relay via a transistor driver, it is possible to turn devices on and off automatically. If the 10th output (pin 11) is connected to the reset pin (pin 15) the sequencer will automatically recycle. In addition to testing various devices, this sequencer can be used to turn lights in different rooms in your house on and off automatically while you are away.

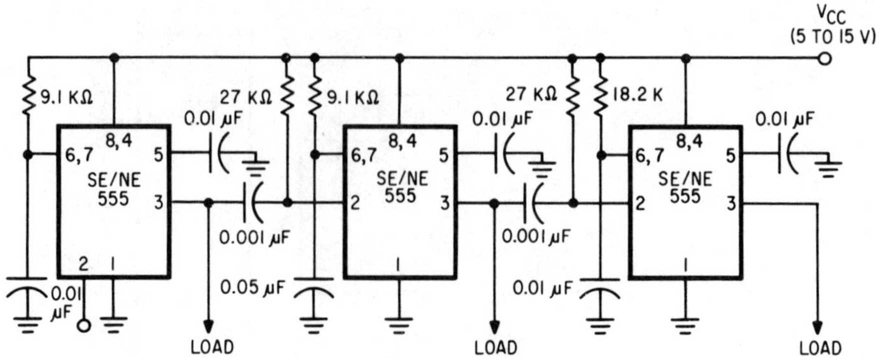

Fig. 5-40.   Improved sequencer.

5.37   improved sequencer

While the sequencer in the previous section is simple and inexpensive to build, it does not permit you to turn different devices on for different periods of time. All devices must go on and off for the same period of time.

But with the circuit in Fig. 5-40, each device can be turned on in sequence for different periods of time, if desired. The reason for this is that each load has its own timer and the period of each timer can be programmed by selecting the proper timing resistor and capacitor.

To start the timing sequence, it is only necessary to trigger the first timer by momentarily connecting its trigger input (pin 2) to ground. This timer runs for 10 msec, during which time it supplies 200 mA of current to its load. At the end of 10 msec, the output of timer 1 goes low again and triggers the second timer. This timer runs for 50 msec and triggers the third timer.

Of course, the amount of time that each 555 is on can be a lot longer than the short periods selected for demonstration here. If you want the whole sequence recycled automatically, simply couple the output of the last timer in the chain to the first timer.

chapter
six

automotive
applications

6.1 electronic ignition system*

A capacitive-discharge automobile ignition system can be built
with commonly available components. The system (Fig. 6-1) employs
a 555 timer, which operates in an asynchronous square-wave mode, to
drive the system's converter section. Thus, a common 6.3-V center-tap
filament transformer of good quality can be used as the converter
transformer. The rectified output of the converter transformer charges
C2 to approximately 500 V dc.

When the points open, a positive-voltage pulse is coupled through
R10, CR6, and C4 to the gate of the 2N4444 SCR. When the SCR fires,
C2 discharges through the spark coil and starts to recharge with the
opposite polarity. This polarity reversal provides a negative charge
through R8 and CR8 to the SCR gate to prevent its retriggering after
the SCR turns off.

When the points close, they discharge C4 through R9 and R10 so
the SCR can be retriggered. The time required for this discharge provides delay to prevent erratic SCR firing caused by point bounce at
high engine rpm.

This circuit is in actual use and has been bench-tested to an
equivalent of 15,000 rpm on an eight-cylinder engine. With careful
shopping, the entire system can be built for less than $15.

6.2 voltage regulator†

A 555-type IC timer, in combination with a power Darlington
transistor pair, can provide low-cost automotive voltage regulation.
Such a regulator can even make it easier to start a car in cold weather.

---

* Morgan, L. G., "Electronic Ignition System Uses Standard Components," *Electronic Design*, Nov. 22, 1974, p. 198.

† Fusar, T. J., "IC Timer Makes Economical Automobile Voltage Regulator," re-
printed from *Electronics*, Feb. 21, 1974; copyright © McGraw-Hill Inc., 1974. All
rights reserved.

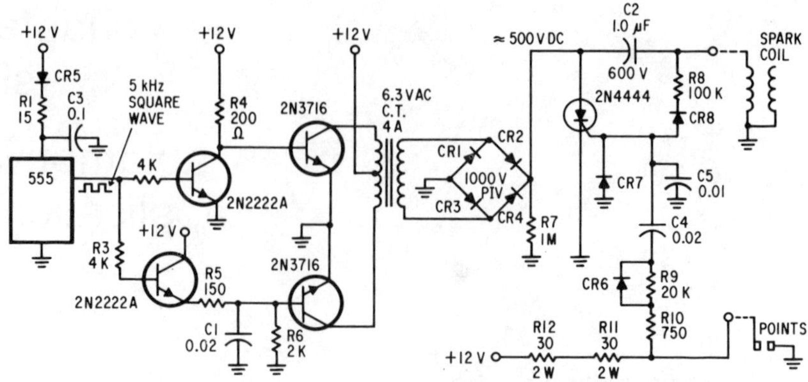

Fig. 6-1. Electronic ignition system.

As Fig. 6-2 shows, the circuit requires very few parts. The value of resistor R1 is chosen to prevent the timer's quiescent current, when the timer is off (output, pin 3, low), from turning on the Darlington pair.

If battery voltage becomes too low, the timer turns on, driving its output high and drawing a current of about 60 mA through resistor R2. This causes a sufficient biasing voltage to be developed across resistor R1 and the Darlington turns on supplying the energizing current to the field coil of the car's alternator. Diode D1 suppresses the reverse voltage of the field coil when the Darlington pair is turned off.

The regulator's low-voltage turnon point is fixed by setting the voltage at the timer's trigger input (pin 2) to approximately half the reference voltage existing at its control-voltage input (pin 5). The high-

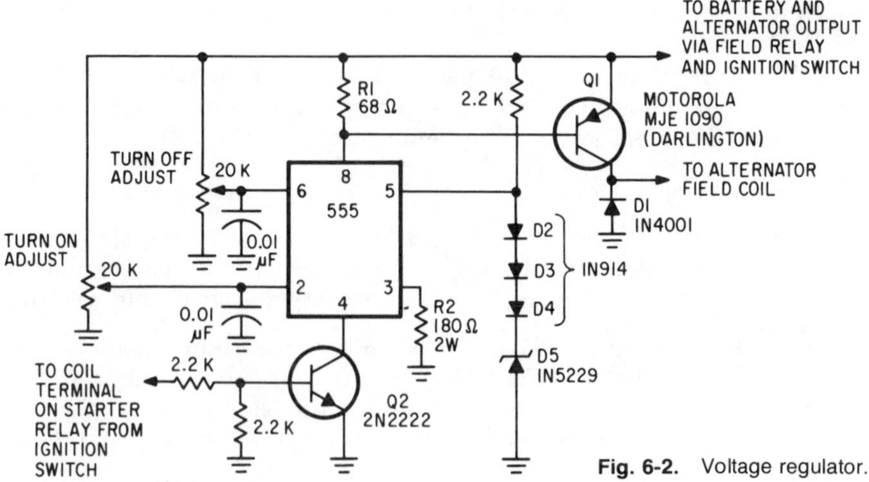

Fig. 6-2. Voltage regulator.

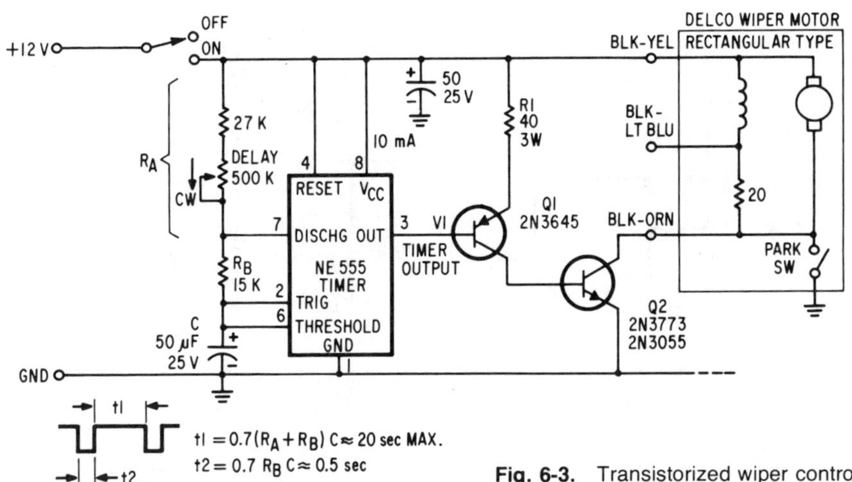

Fig. 6-3. Transistorized wiper control.

voltage turnoff point is set by making the voltage at the timer's threshold input (pin 6) equal to the reference voltage at pin 6. At 77°F, the turnon voltage is typically 14.4 V, and the turnoff voltage is typically 14.9 V. These voltage levels, of course, should be set to match the charging requirement of a given car's specific battery-alternator combination.

The value of the reference voltage is established by the diode string D2 through D5; here, it is approximately 5.9 V. The output voltage has a negative temperature coefficient of -11 mV/°F.

A transistor and a couple of resistors can be added to the circuit for better cold-weather starting. During starting, the transistor holds the timer in its off state lightening the load on the car's cranking motor. (And to prevent radio interference, a 10-μF capacitor can be connected from the Darlington emitter to ground.)

6.3   transistorized wiper control*

An all-solid-state automobile wiper-control circuit allows the windshield wiper to sweep at selected frequencies from once a second to once every 20 sec. The circuit (Fig. 6-3) uses one IC, two silicon transistors, and seven discrete components.

Circuit timing is determined by a 555-timer IC and its external parts, R_A, R_B, and C. Transistor Q1 is switched on when V1 goes low, and npn transistor Q2 also turns on. The mechanical park switch takes over and conducts the motor current until one cycle of wiper motion is complete. At wiper park, the park switch opens and stops the wiper.

* Galluzzi, P., "Circuit Provides Slow Auto-Wiper Cycling with One to 20 Seconds Between Sweeps," *Electronic Design*, Dec. 26, 1974, p. 108.

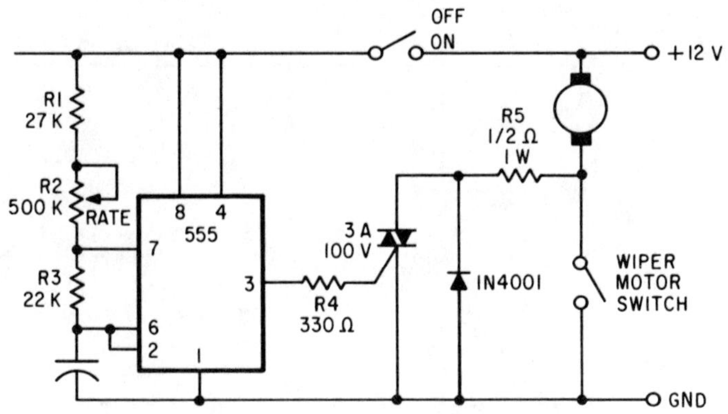

Fig. 6-4.   Thyristor-switched wiper control.

Transistors Q1 and Q2 conduct for only about 0.5 sec. They do not conduct again until the next timer pulse. The delay between pulses is adjusted with the 500-kΩ delay resistor.

Resistor R1 limits the current into Q1 and the base of Q2. The peak collector current into Q2 is about 3 A. Since the duty cycle is normally very low, little heating occurs.

This circuit is in use on a GM-Delco rectangular-motor wiper system.

6.4   thyristor-switched wiper control

As in the previous circuit, the delay in this unit is adjustable from about 1–20 sec. The major difference between this wiper control (Fig. 6-4) and the earlier one is that this one uses a thyristor to do the switching. Like circuit 6.3, it is meant for cars in which the switch for the wiper motor breaks a connection to ground.

Diode D1 (1N4001) is included to prevent the back emf that is produced when the wiper opens at the end of a cycle from retriggering the thyristor and switching it on again without waiting for the delay. The diode can do this because it has a zener breakdown that is lower than that of the thyristor.

The addition of resistor R5 is to ensure that the current through the thyristor falls enough for it to switch off when the wiper contact closes. It may be necessary to increase the value of it a bit if this does not happen.

6.5   relay-switched wiper control

This wiper control (Fig. 6-5) is a more deluxe version of the two preceding ones. It uses a relay to perform the switching for the wiper and is meant for wiper motors whose switch breaks a connection to the

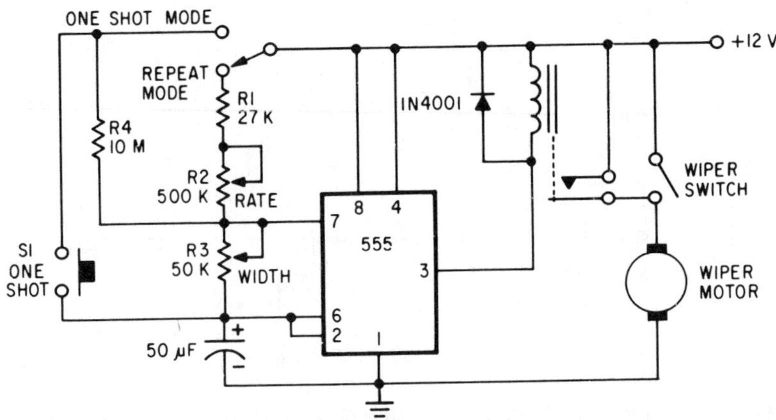

Fig. 6-5.   Relay-switched wiper control.

positive supply rail (that could be changed by simply connecting the relay contacts to another spot).

The 555 astable drives a relay with a frequency that is adjustable by R2. A feature of this unit, which was not on the others, is that it has a variable-width control, so that the amount of time that the relay is on can be adjusted.

Another feature of this wiper control is that it offers two modes of operation: the normal cyclical mode and a one-shot mode. In the one-shot mode, the wipers can be activated for one cycle by pressing button S1 momentarily. If the button is not pressed again, in about 5.5 min the unit will itself activate the wipers for one cycle. This can serve as a reminder that it is still on.

6.6   seat-belt alarm

For those of you who like to wear seat belts in the car and have trouble convincing others that they should too, this circuit is ideal. It is an astable multivibrator whose output is connected to a power amplifier and a speaker (Fig. 6-6).

The loud wail that this circuit produces (about 5 W) should convince anyone to put on his seat belt, because that's the only way to stop it. It works like this: a magnetic reed switch that is normally open when there is no magnet near it is connected to the base of transistor Q1. So is R3. As long as the reed switch is open, R3 supplies current to the base of Q1 and turns it on. Q1 in turn permits current to flow to the astable circuit and the unit screams.

As soon as a magnet is brought near the reed switch, its contacts close, R3 is shorted to ground, Q1 turns off, and the oscillator turns off. All this takes place only if S2 is on, which occurs when someone sits in the seat. The reed switch and the magnet should be glued or

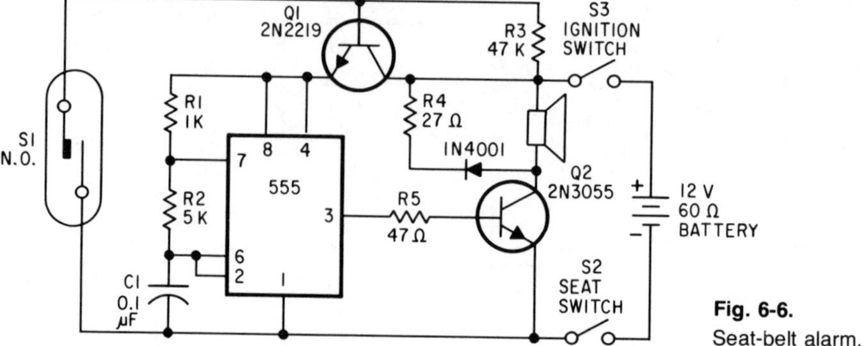

Fig. 6-6.
Seat-belt alarm.

taped to the seat-belt buckle in such a way that when the seat belt is properly secured, they are in close proximity to one another.

6.7   seat-belt reminder

This circuit, unlike the previous one, does not force you to wear the seat belt when you are in the car. Rather, it reminds you that you should put it on, but obediently shuts up if you tell it to.

Once again, we see that the astable connection of the 555 is the one that comes in handy. Like the former circuit, this one (Fig. 6-7) uses a power amplifier on the output, to make sure you don't overlook the signal.

The hot lead for the circuit is connected to a point in the electrical system of the car that receives electricity only when the ignition switch is on. In most cars, a connection can be made to the supply lead for the radio. The ground lead for the timer circuit is connected to the anode

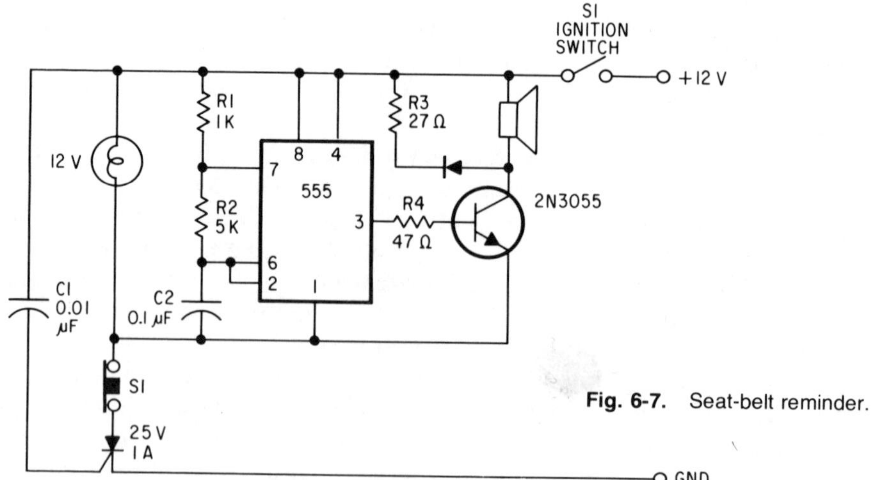

Fig. 6-7.   Seat-belt reminder.

of an SCR via pushbutton S1. As long as the SCR is not triggered, the oscillator will not operate.

However, when the ignition is turned on, a pulse passes through capacitor C1 to the gate of the SCR, because for an instant, the capacitor behaves as a short circuit. The capacitor, however, quickly charges up and will prevent further triggering of the SCR. In the meantime, the SCR has been turned on by the trigger pulse and it acts as a short circuit so that now the astable starts to oscillate. The astable will remain on until the SCR is turned off. This is done by simply pressing on the pushbutton switch for a moment, to break the circuit.

The lamp in the circuit can serve a dual purpose. First, it is there to insure that enough current flows through the SCR so that it will remain in conduction. If the current is too low, as it might be with the astable circuit alone, the SCR would be starved and would not latch. Second, if the lamp is part of the switch assembly for S1, then it will be very easy to locate the shutoff switch at night, when the interior of the car is dark.

6.8   low-battery alarm

What's the condition of your car battery? Is it low? Have you ever checked it? Chances are you cannot answer any of these questions satisfactorily. And if not, then you need this circuit. It is a low-battery indicator that will sound a tone when the voltage on your battery drops below 10 V.

As seen in Fig. 6-8, a zener diode is chosen whose zener voltage is equal to the low-limit voltage of the battery under test.

In this case, it was decided that if the car battery voltage dropped to below 10 V, the alarm should go off. So a 10-V zener was selected. With the zener connected as it is, 10 V is dropped across the zener and 2 V is placed on the junction of R1 and R2. This causes transistor Q1 to conduct, which in turn prevents Q2 from conducting, hence no alarm.

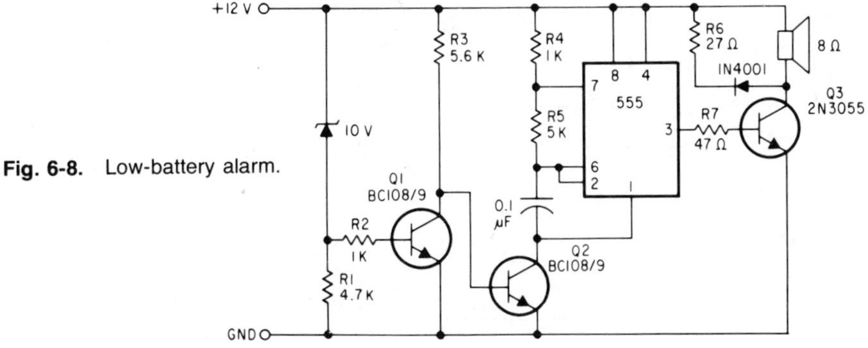

Fig. 6-8.   Low-battery alarm.

However, if the voltage at the input to the circuit drops below 10 V, the zener diode will stop conducting and Q1 will turn off. This will cause Q2 to turn on, and will supply a ground return for the 555 oscillator, resulting in a tone being generated.

6.9 back-up alarm

Backing out of a long driveway can be dangerous, especially if there are small children around who cannot easily be seen. With this little circuit (Fig. 6-9), an audible warning tone will be sounded as soon as you put the car in reverse. The sound will stay on until you take the car out of reverse gear.

Basically, the device is an amplified oscillator whose output is used as the warning signal. For cars that have a separate set of back-up lights that turn on when the car is in reverse, connecting the unit to the car is extremely simple. In that case, the components inside the box are not needed and point A gets connected to the hot lead of the back-up lamp, while points B and C get connected together and are both connected to the chassis of the car (ground). Now whenever the car is put in reverse gear, the alarm, whose speaker should be mounted in the rear of the car so it can be heard, goes off.

But not all cars have separate back-up lights. Some of them turn on the blinker lights when the car is in reverse. In this case, the components in the box are needed and the circuit is constructed exactly as it appears in Fig. 6-9. In this case, points A and D are connected to the right and left rear blinker lights. Point A supplies power to the

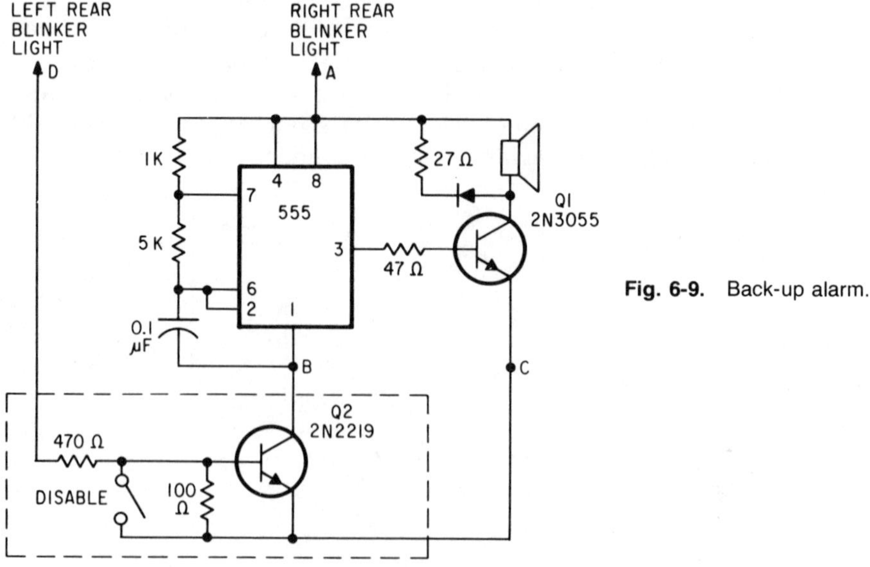

Fig. 6-9. Back-up alarm.

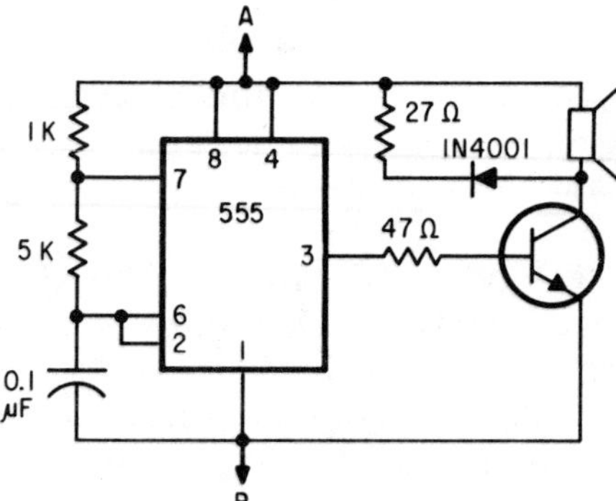

Fig. 6-10. Turn indicator.

oscillator as normal and point D supplies power to the base of transistor Q2. This turns the transistor on and effectively shorts it to ground, causing the oscillator to work.

Remember, this happens only when both of the rear blinker lights are on at the same time. Thus, if your car has a hazard flasher that flashes the front and rear lights together, the back-up alarm will also turn on intermittently with the lights. To prevent this from happening, a disable switch has been included. This switch grounds the base of Q2 and prevents it from turning on.

6.10   turn indicator

Have you ever driven behind a person who had his turn indicator on but goes on for blocks on end without making a turn? It has probably happened to most of us at one time or another. The reason for this is that when the signal is turned on to indicate a lane change, or when one pulls away from the curb, the rotation on the steering wheel is not always enough to cause the mechanical return of the indicator switch. In addition, the clicking sound produced by the flasher inside the car is not always heard.

By using this circuit, which is very similar to the previous one, a loud flashing tone will be produced when the turn signal indicator is turned on, and turned off when the turn indicator goes off.

In the circuit shown in Fig. 6-10, point B is normally connected to the chassis of the car (for negative-ground cars). Point A has to get connected to a point that goes positive for each flash of the turn lights.

In some cars, where there is only one indicator light on the dashboard, it is only necessary to connect point A to the hot side of the light bulb. In most cars, however, there are two turn signal indicators on the dashboard. In that case, point A should be connected to one of the terminals on the flasher module that goes on with each flash.

6.11   headlight extinguisher

An automatic headlight extinguisher (Fig. 6-11) will allow you to turn off the car's ignition and still have a light to open the door by at night. After a predetermined period of time, which can vary from 10 sec to 1 min, the headlights will automatically shut off. Not only does this give you enough light to find your key in the dark, it also prevents you from accidentally leaving the lights on and finding a dead battery in the morning.

It operates like this. When the car ignition is turned on, current flows through resistor R3 and diode D1 to the relay. The relay then pulls in and makes it possible to turn on the headlights. When the ignition switch is turned off, a negative-going pulse is generated and applied to the trigger input of the timer (pin 2). Since the timer is configured to operate in the monostable mode, the pulse causes the output of the timer to go high for a period of time determined by t = 1.1(R1 + R2)C1. In this case, the pulse width is adjustable from about 10 sec to 1 min. The output of the timer is connected to the relay so the relay stays high for the additional period of time after the ignition is turned off.

It should be noted that the headlights will stay on only if the headlight switch is not shut off. In addition, you must remember to turn off the headlight switch the next morning, or you'll be driving around all day with your lights on.

6.12   light alarm

This alarm unit is a handy accessory to use with the headlight extinguisher in the previous section. As in most of the alarm-type

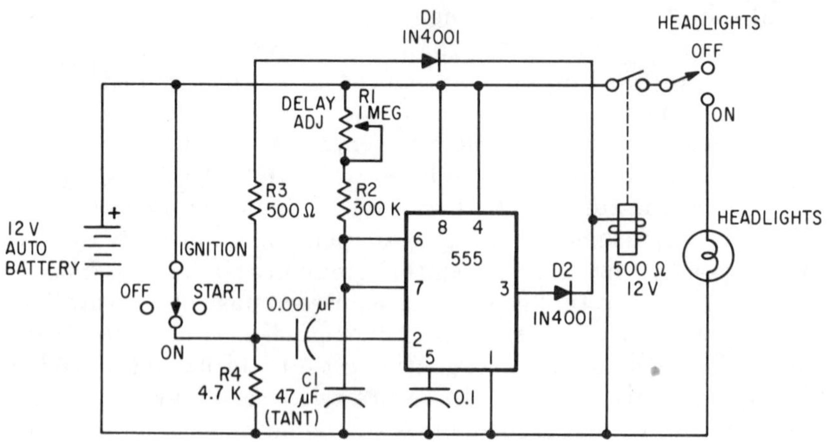

Fig. 6-11.   Headlight extinguisher.

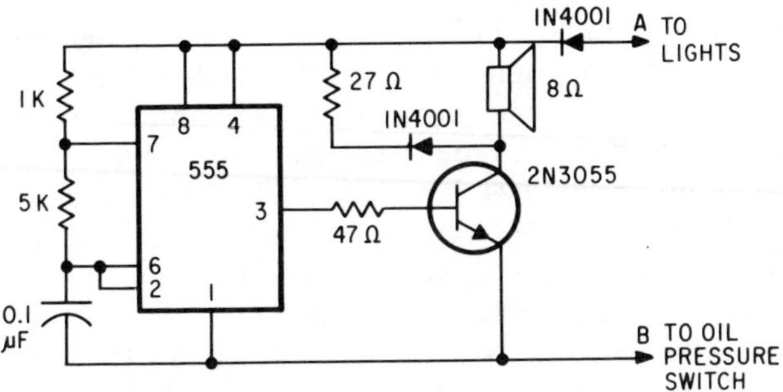

Fig. 6-12. Light alarm.

circuits, this one (Fig. 6-12) is composed chiefly of an amplified astable multivibrator. In addition, there is a diode in the positive power lead to protect the circuit from reverse voltages. Operation is very simple. When the ignition is on and the headlights are on, both point A and point B have +12 V applied to them, and the circuit has zero voltage drop across it so it does not operate.

When the ignition is turned off, however, the oil-pressure switch shorts to ground, and if the headlights are on, they supply power to the oscillator and a warning sound is generated.

6.13   automobile burglar alarm

With car theft on the rise, a good burglar alarm can be a useful thing to have. In Fig. 6-13 is a circuit for a simple alarm that uses a single 555 in the astable mode.

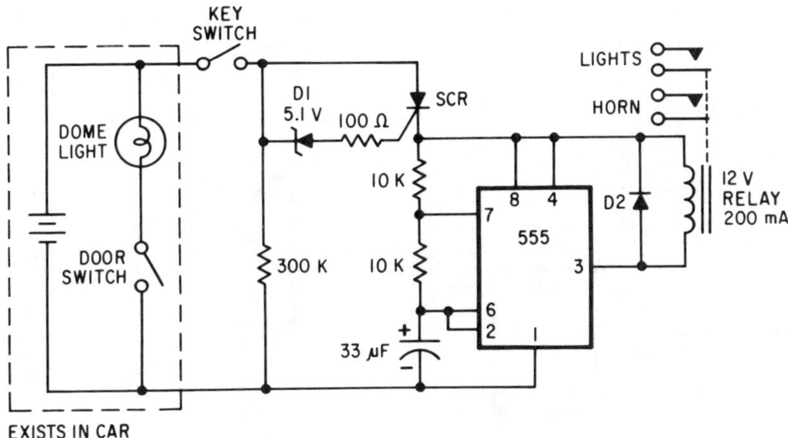

Fig. 6-13.   Automobile burglar alarm.

The alarm is connected to the already existing door switches that turn the dome light on when the door is opened. When the key switch, which is located on the fender of the car, is on, and one of the car doors is opened, a triggering voltage is applied to the gate of the SCR. This turns the SCR on and causes it to latch. The SCR thus applies power to the astable circuit, which oscillates at a frequency of about 1.5 Hz. The output of the oscillator drives a relay. Diode D2 is used to prevent the timer from latching on due to the back emf generated by the relay coil. If a double-pole relay is used, the circuit can turn both the horn and the headlights on and off. The horn blowing will surely scare away any potential thief, and the flashing headlights will indicate to passersby which car is being tampered with.

The SCR is used to latch the circuit on so that, even if the thief closes the car door right away, the alarm will stay on until it is shut off with the key switch.

6.14   keyless burglar alarm

A big disadvantage of the alarm in the previous section is that it requires that a key switch be mounted outside of the protected area, generally on the fender of the car. But by adding another timer, or using a dual timer such as the 556, an alarm circuit can be built that can be armed with a hidden switch that is located somewhere inside the car.

What makes this possible is the second timer, which introduces a time delay before it arms the alarm. As long as you leave the car and close the door before this delay period expires, you'll have no problems.

This circuit (Fig. 6-14) requires that special switches be installed at each door, because it cannot use the existing one. All of the switches must be connected in series and are all normally closed when the doors are shut. The door switches short out the timing capacitor of T2. When one of the doors is opened, the short across C4 is removed and the

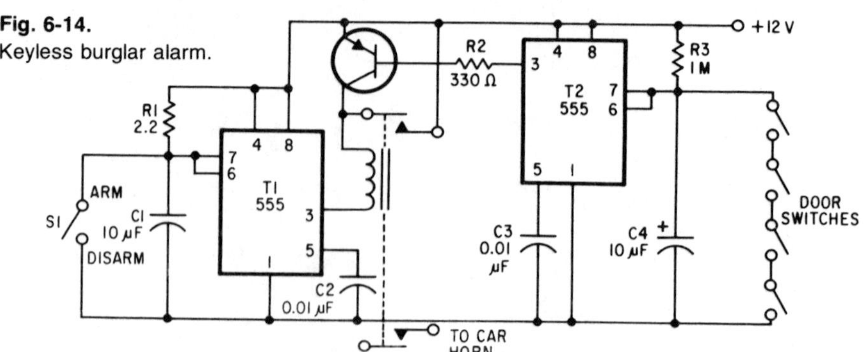

Fig. 6-14.
Keyless burglar alarm.

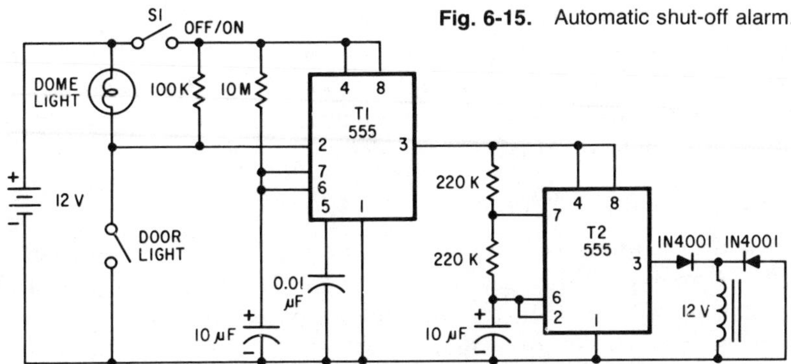

Fig. 6-15. Automatic shut-off alarm.

capacitor starts to charge up. This will take about 11 sec with the components shown. After 11 sec, the output voltage on pin 3 of T2 drops, and causes the transistor to turn on.

If the voltage at the output pin of T1 is low, the relay, driven by the transistor, will close. This does two things. It closes the contacts that are used to operate the car's horn and it also latches the relay on via a second set of contacts. Thus, the relay will remain on, and the horn will sound, as long as the output of T1 is low.

S1, the hidden switch, is used to arm and disarm the alarm and can be hidden somewhere inside the car. When S1 is closed, timing capacitor C2 is shorted and the voltage at the output of T1 is almost at 12 V. Thus, the relay will not close when S1 is in this position. And if the alarm has been triggered, it may be silenced by closing S1.

To set the alarm, S1 is opened. You then have 25 (t = R1C1) sec to close all of the doors before the horn will sound. On returning to the car, you will have 11 sec (t = R3C4) to disarm the unit before the alarm sounds.

6.15 automatic shut-off alarm

A nice feature that neither the two previous alarm circuits has is automatic shutoff. This alarm (Fig. 6-15) uses two timers. T1 is set up·as a monostable, which once triggered provides power to T2 for almost 2 min. T2 is set up as an astable that turns the relay on for 3 sec and off for 1, as long as it gets power from the monostable. After the monostable pulse ends, the alarm shuts off and is ready to be triggered again.

A big advantage of this circuit is that it uses existing door switches. And a key switch isn't absolutely necessary, although it does improve security. Here's how it works. S1 is the arming switch; it can be a key switch or simply hidden somewhere externally on the car. Once the alarm is triggered, it can only be shut off by opening S1. To turn the alarm circuit on, you get out of the car and lock all of the doors. Then turn on S1. Anyone who now opens a door will trigger the alarm, which will stay on for only 2 min unless the door remains open. In that case, the alarm continues to blow the car horn until 2 min after the door is shut or until S1 is opened.

6.16   engine immobilizer

An alarm alone is not sufficient protection from auto theft, especially if an experienced thief is involved. Generally, it's a good idea to have other obstacles in the way of the potential thief. One that is quite effective is an engine immobilizer.

Some immobilizers simply consist of a single-pole, single-throw (SPST) switch that is connected in parallel across the points in the distributor. When the switch is hidden, it does a fair job of making things difficult for a thief. But even they have discovered how to quickly recognize a switch of this type and can disconnect it in a matter of seconds.

But if the idea of an immobilizing switch is combined with a 555 timer, a good antitheft device can result. In Fig. 6-16 is the circuit of just such a device. The 555 is operated in its monostable mode, as a power-up monostable. That means it prevents power from being applied to the load until a certain time period ($t = 1.1R1C1$) has elapsed.

For our immobilizer, the monostable is connected to the 12-V supply via an arming switch, and the ignition switch. If the arming switch is closed and the ignition is turned on, current will flow to the monostable. The instant the timing capacitor starts to charge up, the output of the 555 goes high. Since the relay, which is connected to the timer's high output, is also connected to the positive 12-V supply, there is no voltage drop across the relay and it remains inactivated. Thus, the relay contacts remain open and the engine can be started.

The output of the timer remains high for 30 sec and therefore the car can be started and will run fine, but for only 30 sec. At that point,

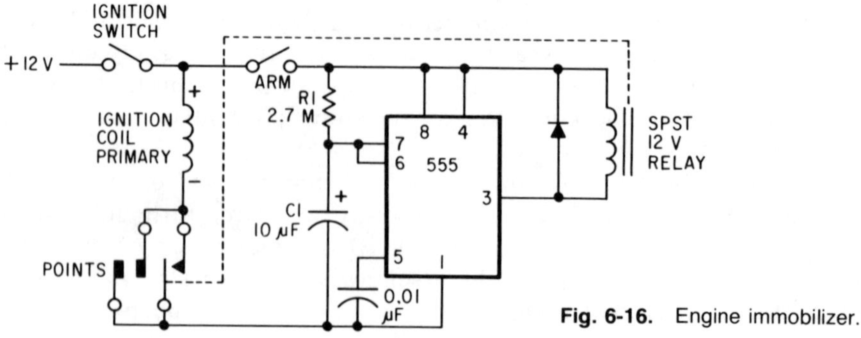

Fig. 6-16.   Engine immobilizer.

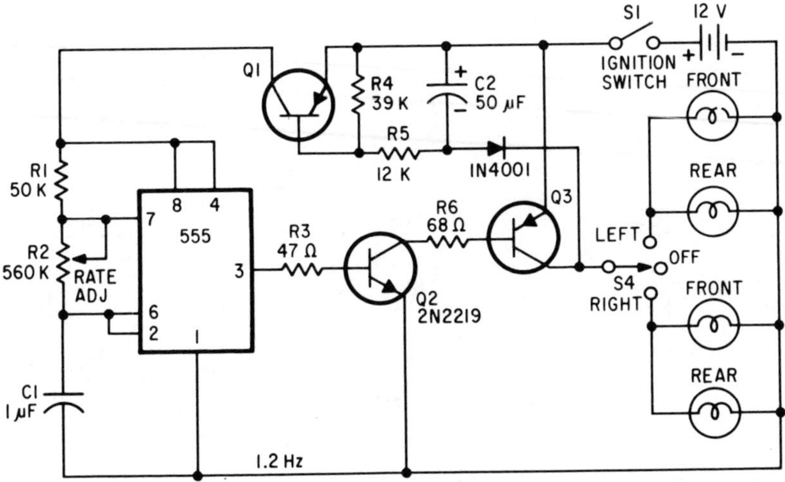

Fig. 6-17. Electronic turn flasher.

the output of the timer will go low again and the relay will turn on, shorting out the points and cutting off the ignition circuit.

If the car is restarted, it will again run for 30 sec and stop. After two or three tries, any thief will abandon this troublesome car for one that is easier to move.

6.17   electronic turn flasher

An all-electronic alternative to the conventional turn-signal flasher is shown in Fig. 6-17. It offers the advantage of having an adjustable flash rate via R2 and overall higher efficiency. The flashing is produced by a 555 astable, but the most important part of the circuit is the circuitry that adapts the one-pole, three-position switch normally found in cars for operation with this circuit, which would ordinarily need a two-pole, three-position switch.

When the turn signal switch S2 is moved from the off position, it permits capacitor C2 to charge via the diode, and the base of transistor Q1 is held on via R5. This turns on Q1 and provides power to the astable. Q3 is prevented from discharging C2 by the diode.

As soon as the direction signal switch is returned to the normally off position, the bulbs stop flashing and shortly thereafter C3 becomes discharged and power is removed from the astable circuit.

6.18   light-up reminder

How often do you ride around in early evening and forget to turn your headlights on? If that happens to you, then this light-up reminder circuit is just what you need. By way of a flashing light in the car, it will tell you when the available light is low enough so that you should

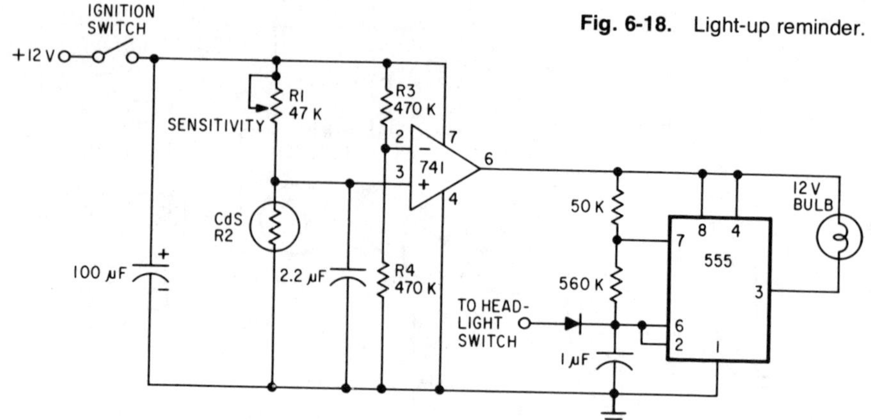

Fig. 6-18.   Light-up reminder.

switch on your headlights. And, if you replace the astable circuit with a monostable and a relay, you can even have it turn the lights on for you automatically.

The circuit in Fig. 6-18 uses an operational amplifier as a comparator in a bridge circuit. R1 and R2 comprise one side of the bridge, while R3 and R4 make up the other. The inverting input of the op-amp is held at half the supply voltage by the R3R4 voltage divider. The voltage at pin 3, the noninverting input, is determined by the R1R2 divider. When the cadmium sulfide (CdS) photocell is brightly lit, its resistance is low and the voltage on pin 3 of the op-amp remains below that of pin 2. Under these conditions, the output of the op-amp will be a voltage that is very close to zero.

When darkness falls, the resistance of the CdS cell increases, thus raising the voltage at the noninverting input. When the voltage reaches the point where it is greater than the voltage on pin 2, the op-amp rapidly amplifies that small positive difference and produces a signal at its output that is close to 12 V. This is the signal that turns the warning circuit on.

The 555 timer is connected to the output pin of the 741 op-amp so that when its output goes high it receives power to cause it to oscillate. The oscillator can be used to drive a warning bulb, flashing it on and off.

Once you turn the lights on, you don't want the flashing light to bother you any more. This problem can be solved in one of two ways. Either you can place the photocell in such a way that it will be able to detect the light produced by the headlights as well as the ambient, or you can sense the voltage that is applied to the headlights.

Sensing the voltage is really quite simple. All that is necessary is to connect a diode to the junction of pins 2 and 6 and the timing capacitor. The anode of the diode gets connected to the light switch, so that when voltage is applied to the headlights it is applied to the anode as well. What this does is to keep the timing capacitor constantly charged, and prevents the 555 from oscillating.

6.19   bad-light indicator

Many times you can drive your car without ever knowing that one or more of your lights isn't working. After all, who checks lights unless it's time to have the car inspected? Not many people, because it means you have to go in the car and turn the lights on then run around the car to make sure they're all working. And if you want to check your brake lights, you have to get another person to help you. One of you has to step on the brakes, while the other checks to see if the lights are on.

Well, checking your car lights can now be as simple as turning a knob. With the circuit in Fig. 6-19, all you have to do to check out all the lights on the car is to sit in the driver's seat and select the proper photocell to connect into the circuit.

In this case, the 555 is being used as a comparator. The photocell array and R1 and R2 compose a voltage divider. If a light is good, it illuminates one of the photocells and the resistance of that cell will drop. This will cause the voltage applied to pins 2 and 6 to rise. R2 is adjusted so that the voltage rise is above $\frac{2}{3}V_{cc}$. When that condition is met, the output of the 555 goes low and the voltage drop across the LED is close to zero. The LED doesn't light.

When a lamp is bad, the resistance of the photocell will be high. This causes the voltage at pins 2 and 6 to drop below $\frac{1}{3}V_{cc}$ and the output of the timer goes high, turning on the LED.

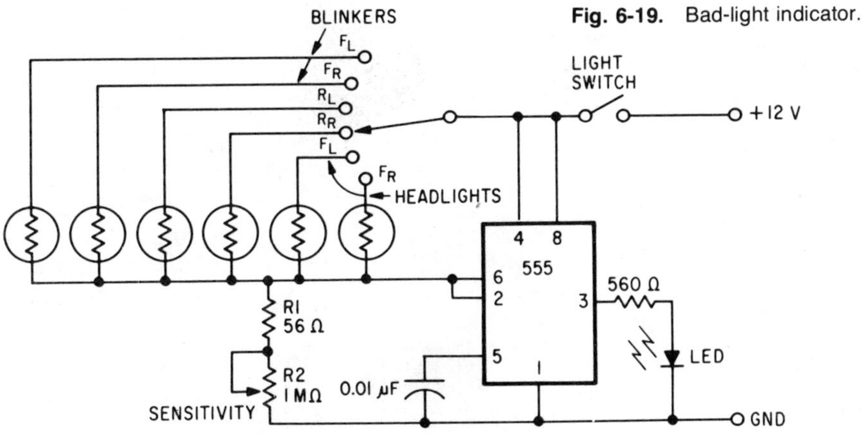

Fig. 6-19.   Bad-light indicator.

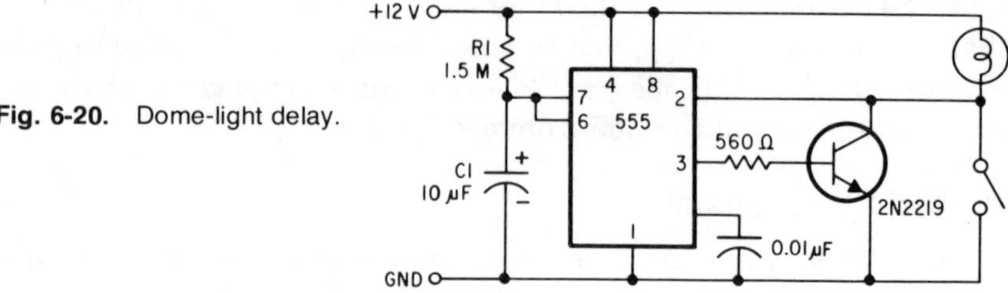

Fig. 6-20. Dome-light delay.

6.20 dome-light delay

By configuring a monostable multivibrator so that it drives the line it senses, you can make a little device that will come in very handy: a delayed-extinguish dome light. This will be useful when you enter your car at night and have to fumble around until you find the ignition keyhole.

The circuit in Fig. 6-20 will keep the dome light on for an additional 15 sec before it turns it off. The time delay is figured out just like it is for a conventional monostable. The output drives a transistor that is connected across the door switch of the car and also to the trigger input.

When the door is opened, the switch shorts to ground and turns the light on. At the same time, it applies a negative spike to the trigger input and starts the monostable cycle. The output of the monostable goes high and stays high for the period t = R1C1. The high output turns on the transistor, which keeps a short across the door switch and keeps the dome light on.

chapter
seven

alarms and
control circuits

7.1 variable-speed motor control*

An IC timer and a few inexpensive components provide an accurate variable-speed control for small synchronous motors. The circuit (Fig. 7-1) can operate from a 12-V car battery or from the 120-V ac line. And, it permits an adjustable frequency range of from 35 to 86 Hz. Since the speed of a synchronous motor is proportional to the frequency of the voltage applied to it, this circuit can act as a controller (Fig. 5-1).

Transistor Q1 and diode D5 comprise a conventional series-pass voltage regulator, which supplies 5 V to the TTL flip-flop (IC2) and the 555 timer IC1. The timer operates in the astable mode with a frequency given by:

$$f = \frac{1.44}{[(R3 + R4) + 2(R2 + R1)]\,C}$$

The nominal frequency of the circuit shown is 120 Hz. Resistors R2 and R3 are used to provide coarse and fine frequency adjustments, respectively.

The output of the 555 is then fed to a flip-flop, where the frequency is divided by two and a perfectly symmetrical square wave results. The complementary outputs of the flip-flop are then used to drive a push–pull output stage that increases the power-handling capability of the circuit. This stage is composed of transistors Q2 and Q3 which are 2N2219 devices, and Q4 and Q5, which are 2N3055s. The filament transformer, which is hooked up backward, couples the circuit output to the motor. A 2-μF capacitor is connected in parallel across the motor to remove some of the high-frequency components that result from the square-wave driving signal.

With the components shown, the frequency deviation is less than 0.3% for dc inputs of 10.5–15 V, or ac inputs of 95–125 V. The maximum

* Hefner, R. D., "Variable-Speed, Synchronous-Motor Control Operates from 12-V Battery or 120-V AC Line," *Electronic Design*, Sept. 27, 1973, p. 110.

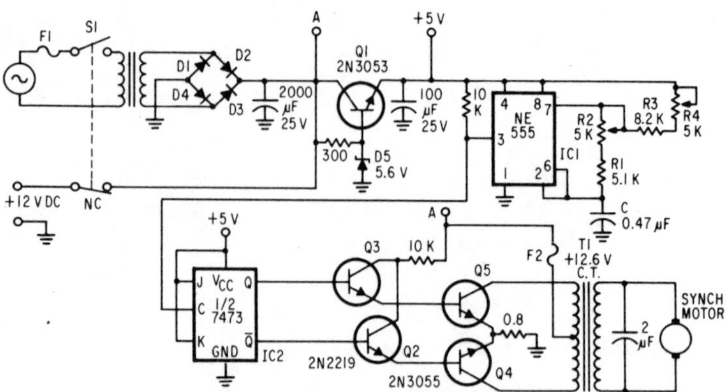

Fig. 7-1. Variable-speed motor control.

motor load is 15 W, a limit determined by the transformer primary current and the IR drop in the emitter-resistor of Q4 and Q5.

7.2   temperature controller*

The internal trigger comparator of the 555 can be used with a thermistor-resistor divider to build an inexpensive temperature controller (Fig. 7-2). You don't even need a well-regulated supply, since all the voltages are derived as ratios of V_{cc}.

When thermistor R3 cools below a set value, which is determined by R2, the voltage at pin 2 (trigger) of the 555 drops below $\frac{1}{3} V_{cc}$. This causes the timer to go high and turns on the triac-controlled heating element R_L, starting the timing cycle. If the thermistor temperature rises above the set point before the end of the timing cycle, the heater shuts off at the end of the timing cycle; otherwise it stays on until the thermistor reaches that value.

Thermistors of different values can be used as long as the relationship R3 + R2 = 2R1 holds true at the desired temperature. Larger values of R2 provide wide adjustment ranges, but sensitivity is reduced.

A timing interval of 1.1R4C is selected to be small, compared with the thermal time constant of the system. However, it must be long enough so that it will prevent the excessive rfi that is caused by the triac switching on and off. It is important to note that the thermistor must be in the vicinity of the heating element for it to work properly.

7.3   light controller

The 555 can be used to automatically turn things on and off as the intensity of light changes. In this application, it is necessary to

* Lewis, G. R., "Low-Cost Temperature Controller Built with Timer Circuit," *Electronic Design*, Aug. 16, 1975, p. 82.

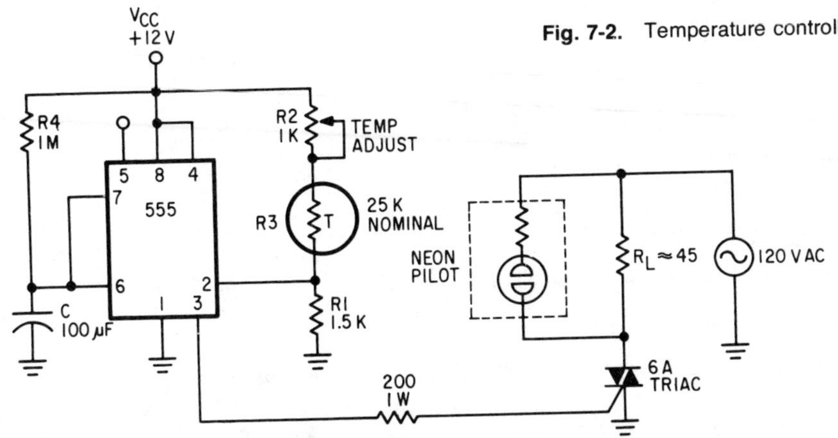

Fig. 7-2. Temperature controller.

use an element whose resistance will vary with the amount of light projected onto it. A light-dependent resistor (LDR) or a CdS photocell will do nicely. This light-sensitive element is connected to the timer as illustrated in Fig. 7-3. When there is light on the LDR or CdS cell, the resistance on the device drops to a low level. This changes the effect of the voltage divider and raises the voltage on the threshold input (pin 6). When the resistor ratios change to the point where the voltage on pin 6 rises to $\frac{2}{3}V_{cc}$, the output of the timer goes low and the relay turns off.

When the ambient light around the CdS cell decreases, the resistance of the cell increases and the voltage at the trigger input to the 555 decreases. When the voltage drops to $\frac{1}{3}V_{cc}$, the timer triggers and causes the output to go high.

To calculate the values of R1, R2, and R3, we must first know the ratio of the resistance of the photoresistor when it is dark and when it is light. This ratio $R_d/R1 = a$. If "a" is equal to or greater than 2, then the resistors can be figured out as follows: $R1 = aR_d$; $R2 = (0.5a - 1)R_d$; $R3 = (3a^2 - 1)R_d/(4a - 2)$. However, if the ratio "a" is less than 2, then the resistors can be calculated as follows: $R1 = 2R_d$; $R2 = 0$; $R3 = 2R_dR1/(2R_d - R1)$.

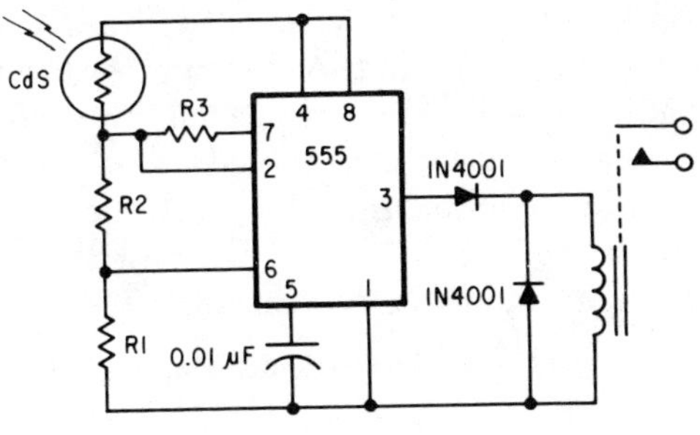

Fig. 7-3. Light controller.

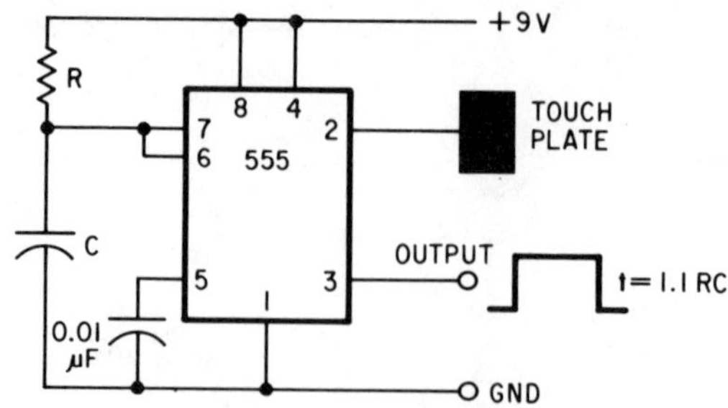

Fig. 7-4. Touch switch.

7.4   touch switch*

A versatile touch switch (Fig. 7-4) can be constructed from the 555 timer and just a few additional components.

Some of the virtually unlimited applications of the touch switch include: switchless keyboards, thief annunicators, activators for the physically handicapped, bounceless electronic switches (with no moving parts), and novelty controls.

The timer itself features either free-running or one-shot capabilities, which can be controlled through the use of "trigger" and "reset" inputs. The characteristics of the output pulses are fully adjustable over a large duty cycle, with timing periods adjustable from microseconds to hours. The output is capable of sinking 100 mA for either electromechanical activation or other interfacing applications. Supply voltages are noncritical, since the device is specified for operation between 4.5 and 16 V dc. For 5 V operation, the device is directly TTL compatible and draws only 3 mA, making it suitable for battery operation. At 16 V, the timer draws on the order of 8 mA.

The trigger input on the device is the key feature in touch-switch applications. Requiring only 500 nA to fire at ⅓ the V_{cc} supply voltage (referenced to circuit ground), the device is easily triggered by the voltage differential found between a floating (nongrounded) human body and the circuit itself. This is 20 V or more, depending on static buildup. The touchplate can be any conducting material with virtually no size limitations.

Once triggered, the device cannot be retriggered and it will time out. However, if the duration of human contact exceeds the RC time constant of the timer, random spikes occur in the output after the time out. This can be avoided by making fairly large time constants, so that the device will not time out before contact is removed.

The duration of the output pulse is controlled by both the RC time constant and the control voltage input (pin 5). By varying the voltage at pin 5 the timing can be changed by about one decade. If the entire

* Heater, J. C., "Monolithic Timer Makes Convenient Touch Switch," *EDN*, Dec. 1, 1972.

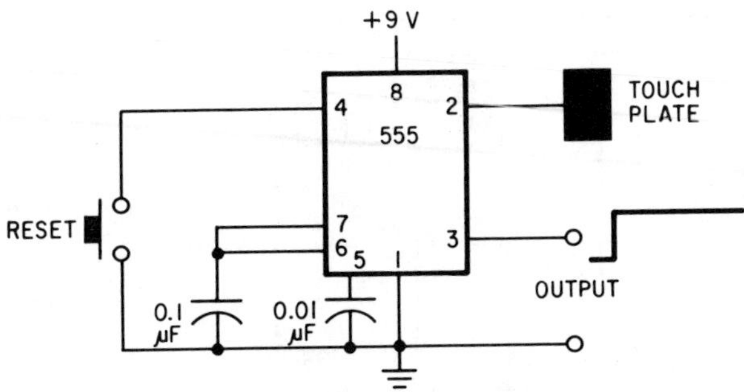

Fig. 7-5.   Latching touch switch.

RC network is omitted and pin 7 is connected directly to pin 6, the circuit will latch "on" when "touched."

7.5   latching touch switch

A modification of the touch switch in the previous section is to make the timing resistor infinite (eliminate it altogether). This was done in Fig. 7-5. The consequence of making the timing resistance so high is that when you trigger the monostable by touching the plate, the output goes high and stays there. In other words, the timer has latched.

In order to restore the 555 to the previous state, where the output is low, you have to reset the device. This is done by momentarily grounding the reset pin (pin 4).

7.6   touch-on, touch-off switch

Circuit 7.4 can be used as the basis of a touch-on, touch-off switch. As can be seen in Fig. 7-6, all that you have to do to get this alternate action switching is to add a flip-flop to the circuit.

In this case, we have added a 7473 J-K flip-flop. To ensure proper operation, the output of the 555 is fed to the clock (pin 1) input and the J and K inputs are tied together and to the positive supply rail. Remember, if you are going to use this circuit you must use a 5-V supply. If that is an inconvenience, then the CMOS 4027 flip-flop can be used. Again, remember to tie the J and K inputs to the positive supply rail.

7.7   ultrasonic controller*

A signal from an ultrasonic transmitter, beamed to a transducer in a receiver unit up to 20 ft away, can be used as a controller, an

---

* Dance, J. B., "Ultrasonic Transmitter/Receiver Generates a 20 Ft. Beam That Detects Objects," *Electronic Design*, Aug. 2, 1975.

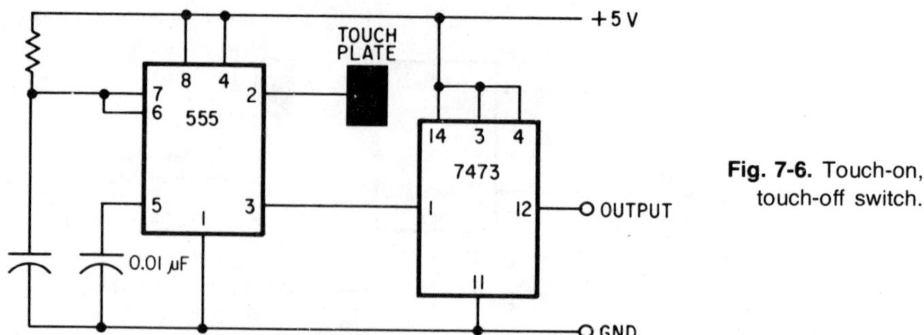

Fig. 7-6. Touch-on, touch-off switch.

invisible antiintrusion device, an object sensor on industrial conveyor-belt systems, a person detector on automatic door openers, and many other applications that can operate from a relay.

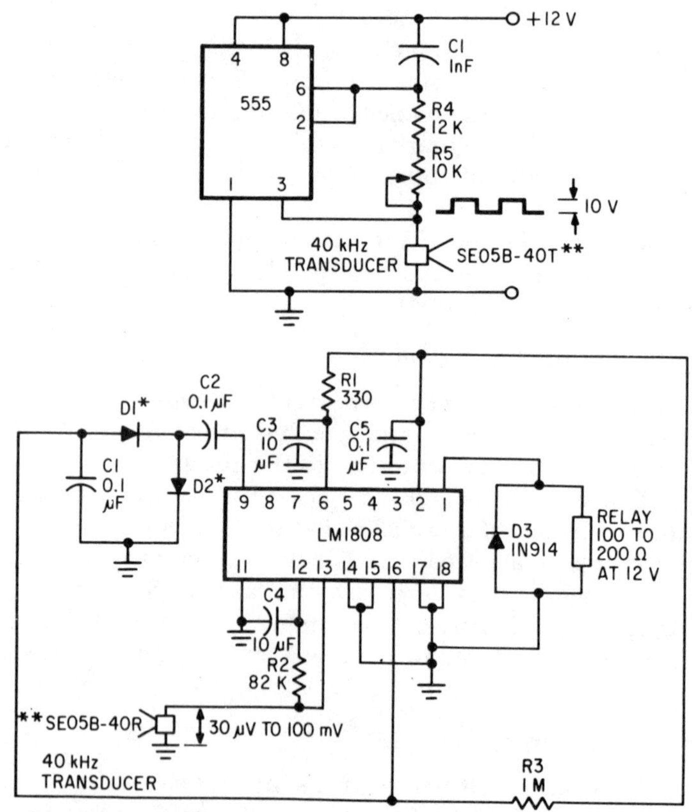

*SEE TEXT

**AVAILABLE FROM HALL ELECTRONICS, AVONDALE RD., LEYTON, LONDON, E17 8JG

Fig. 7-7. Ultrasonic controller.

Most circuits for ultrasonic generation and detection are complex and need high-gain stages to amplify weak received signals, but the circuit in Fig. 7-7 is extremely simple. It uses only two ICs.

The transmitter has a 555 timer, which is connected in an astable mode to provide a squarewave output of approximately 50% duty cycle. The frequency is adjusted by R5 for optimum sensitivity with the transducers used. The values shown are for operation at about 40 kHz.

In the receiver, the LM 1808 IC (National Semiconductor) controls the relay directly. The LM 1808 is equivalent to combining an LM3065 IF amplifier with an LM380 power amplifier but in a single 18-pin DIP. The volume-control section of the LM1808 is not used in this application. And although the 1808 is designed for 24 V, in this application a 15-V supply is adequate.

The output from the 40-kHz receiver transducer enters pin 13 of the 1808, which is the input terminal of emitter-coupled, cascaded differential IF amplifiers. Resistor R2 maintains pin 13 at about the same potential as pin 12, and resistor R1 reduces the potential at pin 6 to about 11 V, which is stabilized by the internal circuit.

The output from the IF amplifier at pin 9 is a square wave when the input at pin 13 exceeds the threshold voltage. Diodes D1 and D2 are low-forward-drop germanium diodes connected in a "diode-pump" circuit to supply a dc negative output when the receiver transducer picks up the 40-kHz beam.

A bias current via R3 to the inverting input of the 1808 power stage at pin 16 results in a 2-V quiescent level at its pin-1 output. When a 40-kHz signal appears at pin 9, the output from the diode circuit causes the voltage at pin 1 to rise to about 13 V, and the relay closes. Diode D3 suppresses the relay coil's current turnoff surge voltage. The relay should deenergize when the coil voltage falls to about 2 V, unless a latching circuit is required. A relay with a lower release voltage may be used if a low-voltage zener diode or about three forward-biased silicon diodes in series are included in the pin-1 circuit to reduce the voltage across the relay coil.

In the author's application, the relay was outfitted with 10-A DPDT contacts.

The transmitter's 40-kHz transducer has an impedance of about 200 Ω and the receiver a value of about 70 kΩ. In both units shown in the figure, the capacitance is listed as 1,400 pF $\pm 20\%$. The receiver unit is sharply resonant, but the 8.2-kΩ resistor across it reduces the Q so the transmitter frequency is not critical. Completely different transducer types, with the transmitter and receiver transducers both identical (400-Ω impedance, 1,700 pF), provide very similar performance. Thus, the choice of transducers is not critical.

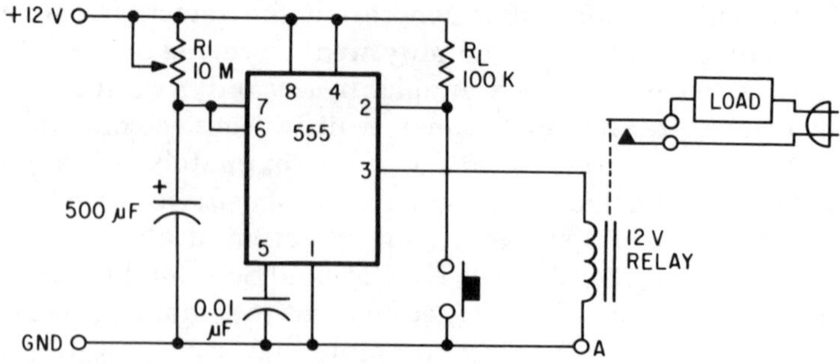

Fig. 7-8.   Delayed shutoff.

7.8   delayed shutoff

By using the 555 as a long-period monostable multivibrator, you can make a delayed shutoff switch that will turn something off 1.5 hr after the time you turn it on.

The design of the unit (Fig. 7-8) is fairly straightforward. Triggering is accomplished via a pushbutton that momentarily grounds the trigger input. The output of the 555 is fed to a relay, which in turn controls the device in question.

By simply changing the connection of one wire, this device can be converted into a delayed on controller. In this case, point A should be connected to the positive supply rail instead of ground. Now, when the pushbutton is pressed, the relay will receive +12 V from the rail and +12 V from the output of the timer. The zero voltage drop across the relay prevents it from turning on. With the components shown, turnon can be delayed for as long as 1.5 hr. The time can be shortened, however, by adjusting R1.

7.9   water alarm

This unit has been designed to sound an alarm when the two probes come in contact with water. It can be used to let you know when

Fig. 7-9.   Water alarm.

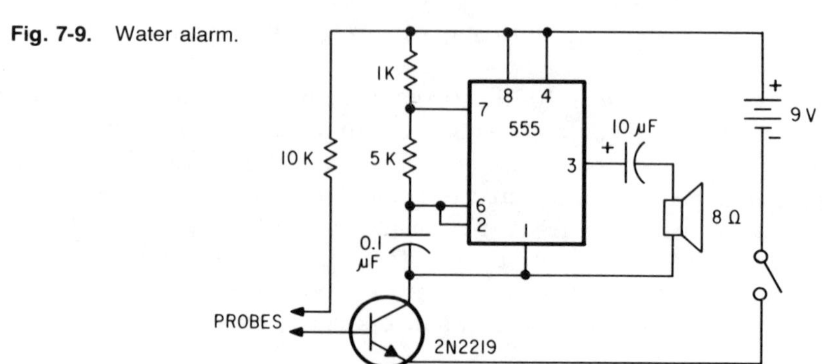

it suddenly starts to rain outside, if the basement has a leak, or as an overflow detector.

As can be seen in Fig. 7-9, the design is rather simple. An astable multivibrator is used to generate the warning signal. To turn the oscillator on, we take advantage of the fact that most of the water we have around us is not pure. It almost always contains some impurities. These impurities convert water, which in the pure state is a nonconductor, into a conducting medium.

When the two probes are placed in contact with water, a current flows from one to the other via the water. This current turns on the base of the transistor and causes it to conduct. The transistor shorts the ground return portion of the astable to the negative side of the battery, and the oscillator produces a tone.

If a 2N2905 pnp transistor is substituted for the npn transistor used, the probes can be placed in a flower pot or garden and, when the soil is too dry, the alarm will go off.

7.10   gas alarm

If a special type of sensor that changes its resistance with the presence or absence of gas is combined with two timers or a single 556 IC, an inexpensive gas detector can be built.

The sensor is known as the Taguchi Gas Sensor (TGS) and is actually a piece of tin oxide semiconductor material that is molded around a small filament heater (Fig. 7-10). The whole unit is housed in a small wire mesh envelope. In use, the filament is heated and if the sensor is in clean air, the resistance of the element is high, anywhere from 20 to 100 kΩ. However, if some combustible gas is present, the resistance of the element drops drastically.

It is this drop in resistance that we use to trigger our alarm. The resistive element of the TGS is connected between the positive supply

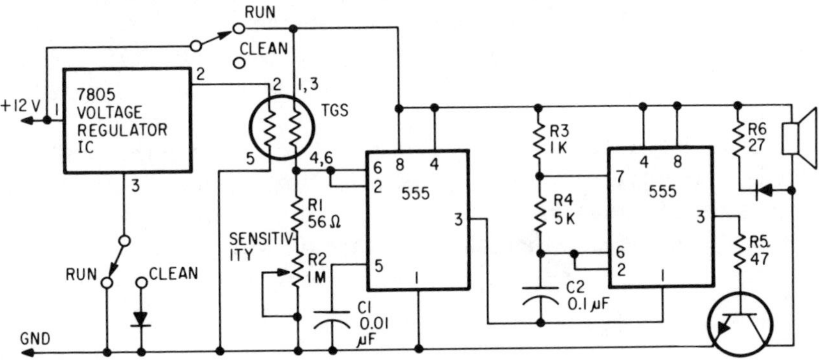

Fig. 7-10.   Gas alarm.

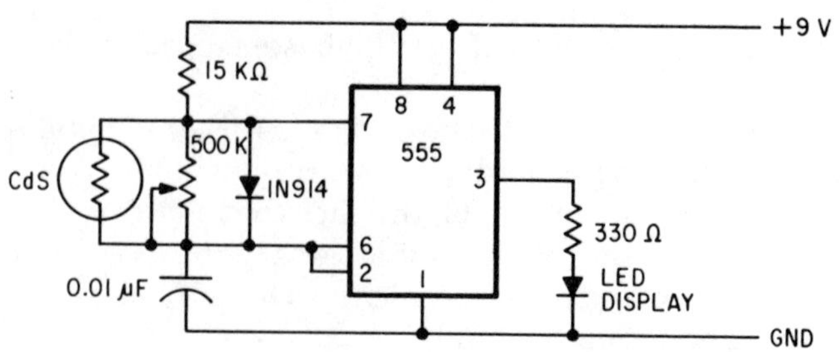

Fig. 7-11. Display-brightness compensator.

rail and the junction of pins 2 and 6 on the 555 (trigger and threshold inputs). The 555 is being used in a comparator mode.

When the resistance of the element drops, the resistance ratio of the sensor resistor to R1 and R2 changes and raises the voltage at pins 2 and 6. R2 is adjusted so that the rise in voltage is above $\frac{2}{3}V_{cc}$. When that happens, the output of the first timer goes low, and timer 2, which is designed to operate as an astable alarm, is activated.

When no gas is present, the resistance of the sensor element is high, the voltage applied by the divider is below $\frac{1}{3}V_{cc}$, and the output of timer 1 is high, disabling the astable.

The sensor filament heater gets its power from a 5-V IC regulator. This is one of the three-terminal devices such as the 7805. A clean/run switch has been included in the unit. The reason for this is that a slightly higher voltage is required to warm the sensor up and to clean it after it has been exposed to gas.

7.11   display-brightness compensator

Automatic adjustment of the brightness of an LED display can be provided by a photocell-timer combination. The timer operates in the astable mode and produces a square-wave signal. The duty cycle of the output waveform is varied by a light-dependent resistor such as a CdS cell. The photoresistor is used to replace what would normally be R2 in the conventional astable circuit. A potentiometer is connected in parallel with the photoresistor in order to adjust the brightness manually.

The circuit as it is in Fig. 7-11 is capable of varying the brightness of an LED display from 10% of full brightness in dark surroundings to 90% of full brightness in bright surroundings. The diode across the photoresistor is used to increase the available duty cycle to more than 50%.

chapter
eight

power supplies
and converters

8.1   transformerless dc-dc converter*

A compact, transformerless dc-dc converter derives a negative supply voltage from a positive one. The technique allows dual-supply op-amps to operate from a single supply line and still deliver bipolar outputs.

A square wave is generated by the 555 timer (Fig. 8-1) plus four external passive components. The nominal period of the resulting square-wave clock signal is given by $T = 0.69 \ (Ra + 2Rb)C$. Five additional passive components are required to derive the negative supply from the clock.

The circuit has component values chosen to give a 2-kHz pulse-repetition frequency, with the coupling and filter capacitors selected to minimize ripple under heavy loads. Since the timer is insensitive to variations in supply voltage and has good output drive capability, it makes an excellent system clock. For a specific application, the capacitor values used depend on required clock prf, load, and ripple rejection.

With a 500-Ω load, typically equivalent to ten 741s, the negative output voltage tracks the positive supply, but its absolute value is always about 3 V lower (Fig. 8-2A). Output regulation for a constant $+10$-V input is approximately 10% for a change from no load to a load of 500 Ω (Fig. 8-2B). Usable outputs are available with load impedances as low as 50 Ω—which would represent about 70 type 741 op-amps.

As an alternative to this approach, you can derive an artificial ground between the real ground and the positive supply, but this would add complexity and reduce the output capability. Of course, you can also use an external supply, but this would increase cost and size.

---

* Pearl, B., "Positive Voltage Changed into Negative and No Transformer is Required," *Electronic Design*, May 24, 1973, p. 164.

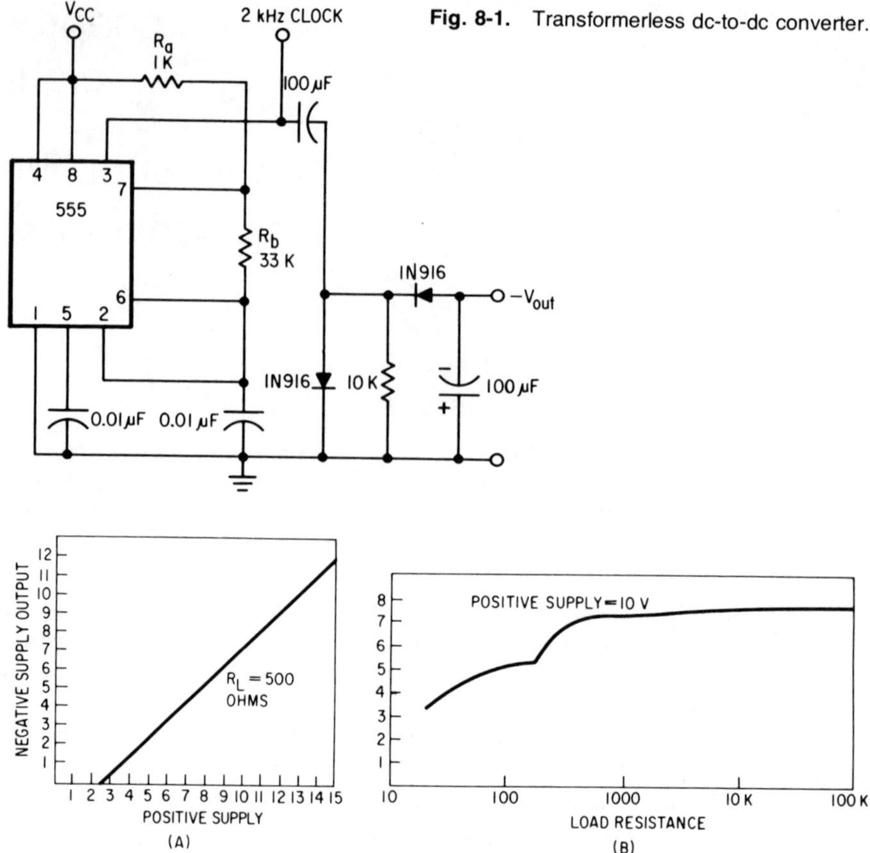

Fig. 8-1.   Transformerless dc-to-dc converter.

Fig. 8-2.   Converter operating curves.

8.2   high-voltage supply*

The addition of a simple feedback-controlled timer circuit converts
a high-voltage supply to a regulated supply (Fig. 8-3). The output
voltage is adjusted with a single control, and the circuit operates from
a +5-V source. The output voltage is directly proportional to C1's charging time, T_{on}, if T_{off}, the discharge time, is held constant. Time T_{on} is
controlled by R1, and T_{off} by R2. Thus,

$$T_{on} \simeq \frac{1.5 V_{out} \, LI_L}{V_c^2} = 0.693 \, (R1 + R2)C1$$

* Johnson, K. R., "High Voltage Power Supply from 5-V Source Regulated by Timer
Feedback Circuit," *Electronic Design*, April 1, 1975, p. 132.

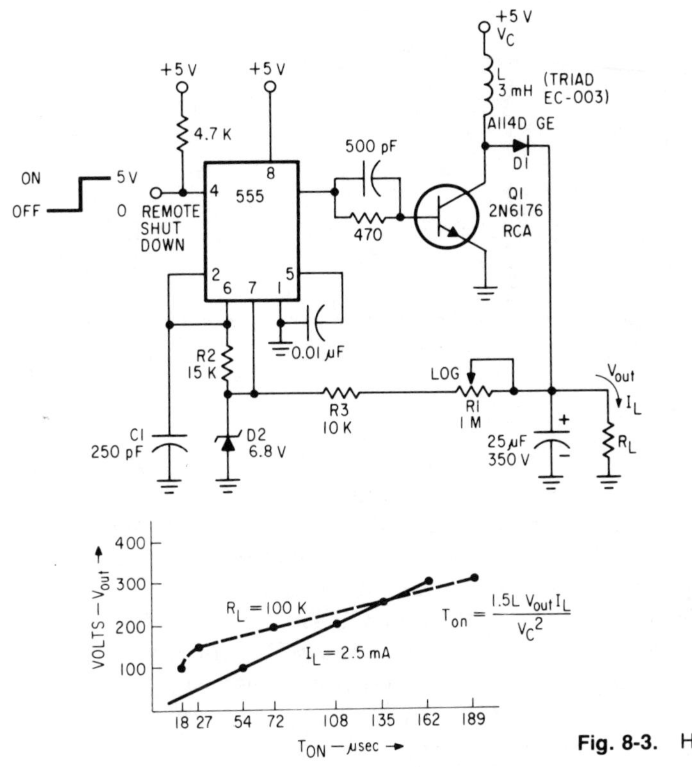

Fig. 8-3. High-voltage supply.

and
$$T_{off} \simeq \left(\frac{V_c}{V_{out}}\right) T_{on} = 0.693 \ R2C1$$

Transistor Q1 is on when C1 charges, and the current increases approximately linearly to store energy in the inductor's magnetic field. The maximum value of current is given by

$$I_L \simeq \frac{V_c - (V_{ce(sat)} \times T_{on})}{L}$$

During T_{off}, Q1 turns off and the collapse of the magnetic field of L generates a voltage, $V_L = L di/dt$, which charges C2 via D1.

Regulation results from the negative-feedback action of V_{out}. A decrease in V_{out} decreases the charging rate of C1 and increases T_{on}. Load regulation is approximately 2%. Output ripple is given by

$$V_r = \frac{I_L T_{on}}{C2} = 12 \ mV$$

for a 2.5-mA load at 250 V. Rectifier-diode D1 should be a fast-switching type, and zener diode D2 protects the timer input.

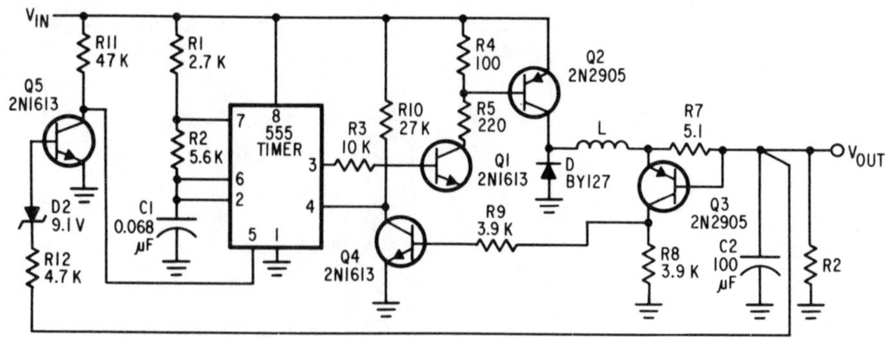

Fig. 8-4.　Switching regulator.

8.3　switching regulator*

Don't despair if you can't find an IC regulator chip for your next design. You can use the 555 timer advantageously for various types of regulators. Among the types in use today, pulse-width-modulated regulators have found great popularity, so the circuit shown (Fig. 8-4) is a PWM configuration complete with current foldback.

For astable operation, R1, R2 and C1 are connected as shown. Capacitor C1 charges to $2V_{in}/3$ through R1 and R2, and discharges to $V_{in}/3$ through R2 when there is no external voltage at terminal 5. Thus, the timer will retrigger itself, yielding a square-wave output with:

$$t_{ON} = 0.695 \ (R1 + R2)C1$$
$$t_{OFF} = 0.695 \ R2C1$$
$$\text{and: } f = 1/(t_{on} + t_{off}).$$

This square wave is amplified by R3, Q1, R4, and R5, and fed to transistor Q2. As long as the timer output is high, Q2 will be on and driving current into R_L and C2 through inductor L. When Q2 turns off, the energy stored in L and C2 is available to supply the load. Differing from the input level, the voltage thus generated is fed to a simple comparator formed by Q5, D_Z, R11, and R12. Q5 will not conduct unless the output voltage is less than the zener voltage. Therefore, the voltage at the collector of Q2 continuously changes depending on how it compares to the zener voltage. Since the collector voltage is fed to the modulating input (terminal 5), the pulse width of the generated square wave is modulated to provide the required output voltage. An approximate relation between V_{in} and V_{out} can be described as: $V_{out} = (t_{on}V_{in})/(t_{on} + t_{off})$.

---

* Chetty, P. R. K., "Put a 555 Timer in Your Next Switching Regulator Design," *EDN*, Jan. 5, 1976, p. 72.

R7 is the current-sensing resistor. When the load current increases to a level such that the voltage drop across R7 turns Q3 on, Q4 will be driven into saturation. The resultant low at pin 4 of the 555 resets the timer, bringing its output to zero. With the timer reset, no voltage develops across R8 and Q4 is turned off, enabling the timer, and bringing its output high. If an overload condition still exists, Q3 and Q4 will again be turned on and reset the timer. This closed-loop chain reaction continues as long as an overload condition exists. If the overload condition increases, the voltage and current will both decrease, initiating the foldback action.

With a 15-V input, the circuit will deliver a 10-V, 100-mA output with line and load regulation figures of 0.5% and 1%, respectively. Foldback action will commence at a current value equal to Q3's $V_{be(sat)}$ divided by R7.

8.4   voltage supply doubler

Many applications require low-current voltage supplies that have a higher voltage than that supplied by the available battery. By using the 555 timer as an oscillator, an 80-mA supply at almost twice the original supply voltage can be produced (Fig. 8-5).

The circuit consists of a 555 timer in a self-triggered mode as a square-wave generator and is followed by a voltage-doubling rectifier circuit. The frequency of oscillation is about 20 kHz and is determined by $f = 0.7/RC$. The 20-kHz rate is particularly convenient because its frequency is so high that filtering is easily accomplished with low-value capacitors.

8.5   three-in-one supply

Here is a useful 555-based power supply that's bound to come in handy. A 555 is used to provide a 20-kHz switching signal (Fig. 8-6). Transistor Q1 matches the output of the 555 to the low-input impedance of the complementary symmetry output stage composed of Q2 and Q3. These transistors operate in a class B switching mode, which improves their efficiency.

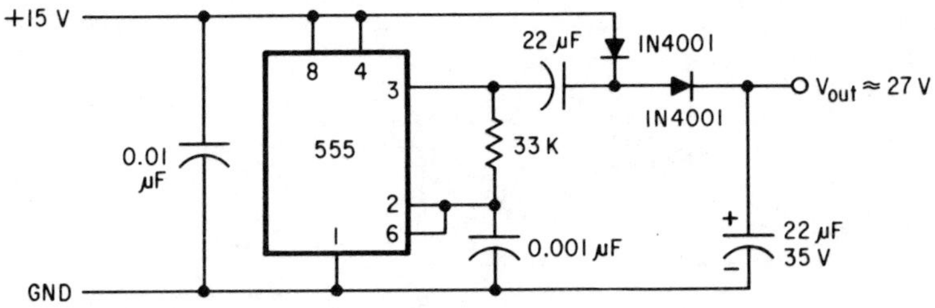

Fig. 8-5.   Voltage supply doubler.

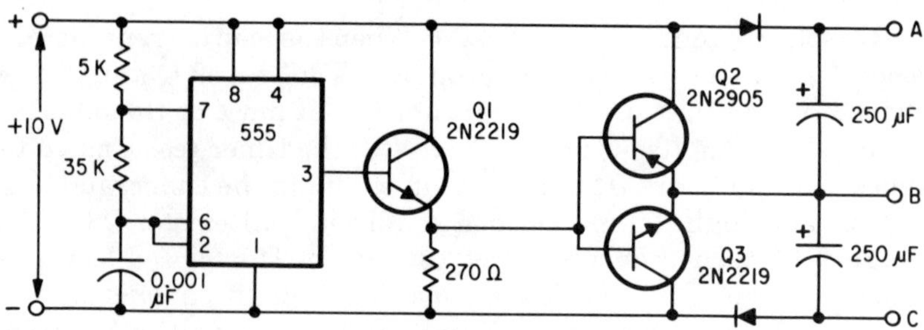

Fig. 8-6.   Three-in-one supply.

Any one of the three output terminals A, B, or C can be used as a common output, thus making it possible to get three different voltage choices from this supply. The available outputs are +20, ±10, and −20 V.

8.6   dc-to-ac inverter

Did you ever wish you could take a 110-V ac power source with you wherever you went? Well now you can. I don't mean for you to go out and buy a $60 inverter. Just build your own with the 555 timer.

The circuit in Fig. 8-7 shows you what you need. The 555 is configured to operate as a square-wave oscillator with almost a 50% duty cycle. The frequency of the oscillator is adjustable from 48 to 68 Hz via R2 so it can be set to provide either 50- or 60-Hz output.

A complementary Darlington output stage provides the drive to the secondary of an inversely connected filament transformer. The output waveform is filtered by the capacitor (C2) and the inductor so that the waveform that reaches the transformer is reasonably sinusoidal.

8.7   improved dc-to-ac inverter

While the inverter in the previous section can be useful in powering ordinary 110-V motors, the frequency may not be that exact. For most applications, that's no problem. However, there are times when

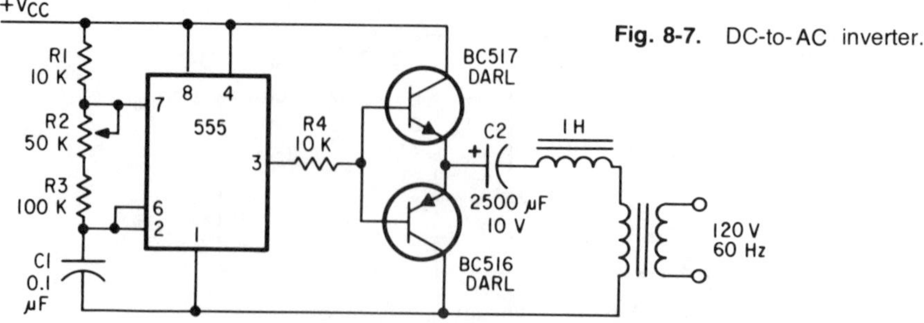

Fig. 8-7.   DC-to-AC inverter.

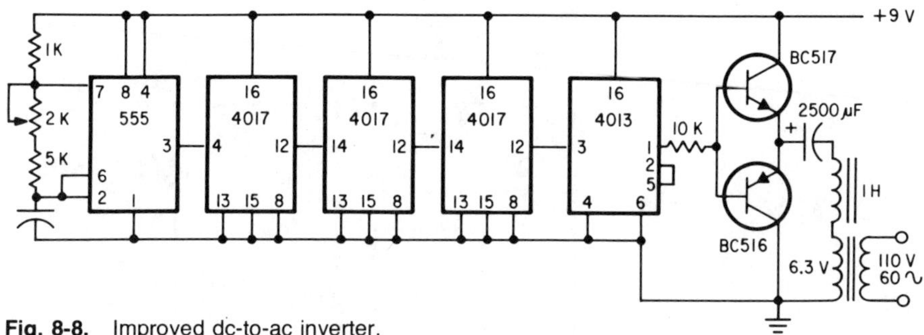

Fig. 8-8. Improved dc-to-ac inverter.

an accurate 60-Hz frequency is a must, such as when you want to run a regular ac clock or a turntable from a phonograph via a battery. Even a slight change in frequency for these applications can result in noticeably deteriorated performance.

A highly accurate 60-Hz inverter can be easily fabricated from the 555 by simply raising the frequency at which it oscillates, and then dividing the output frequency down to 60 Hz. Thus, if there are any slight variations in the oscillator output frequency, they will have a minimal effect on the 60-Hz output.

The timer in Fig. 8-8 is designed to operate at 120 kHz. This is divided down to 120 Hz by three 4017 CMOS divide-by-ten circuits and then divided to 60 Hz by the 4013 flip-flop. The signal is then fed to the same output stage as was used in the previous circuit.

8.8   lead-acid battery charger*

A battery charger for lead-acid cells can be built with a 555 timer at a considerable parts saving over similar circuits. A cell should be charged at a current not exceeding 20% of its rated capacity until a specified terminal voltage is reached. The charging current must then be reduced to approximately 1% of the rated capacity of the battery to maintain the battery in a float condition.

The circuit in Fig. 8-9 meets these charging conditions for a 6-V, 1.2-A system. For this battery, the charge current should be equal to or less than 240 mA and the float-charge current should be approximately 12 mA. When the battery is fully charged, it exhibits a terminal voltage of 7.2 V and a float voltage of 6.8 V.

The two comparators in the 555 detect both the need for charging and the fully charged condition. The timer's internal flip-flop and npn

---

* Kranz, P., "A Simple Battery Charger for Gel Cells Detects Full Charge and Switches to Float," *Electronic Design*, July 19, 1976, p. 120.

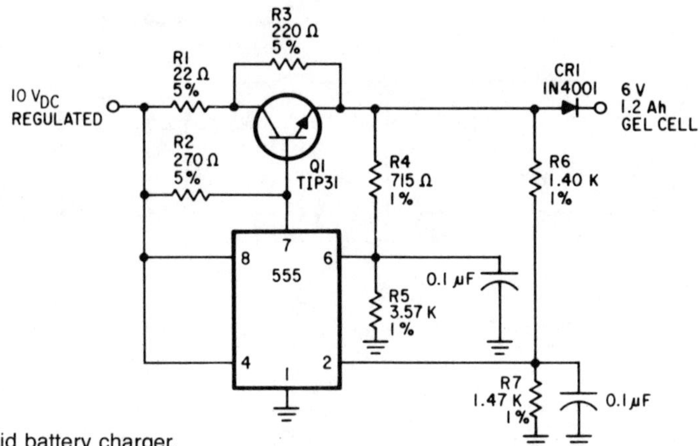

Fig. 8-9. Lead-acid battery charger.

discharging transistor drive an external transistor, Q1. Transistor Q1 switches the charge current from its maximum value (limited by R1) to the float value (limited by R1 + R3). Resistors R4 and R5 set the trip point for detecting a fully charged battery; resistors R6 and R7 set the trip points for detecting a partially discharged battery. Diode D1 disconnects the charging circuit from the battery when the 10-V regulated-input source is removed.

8.9   improved battery charger*

The circuit in Fig. 8-10 is an improved storage-battery charger that uses a 555 timer to "burp" the battery—to apply a 20-V pulse across the battery periodically, while it's being charged. This process tends to shake gas molecules loose from the battery plates. The molecules insulate portions of the battery plates from the electrolyte, which reduces the effective charging current by raising the internal resistance.

This burp-control circuit can be added to just about any existing automobile-battery (12-V) charger. The 555 timer is connected in a free-running mode and has independent on and off times. The time between burps is controlled by the variable pot, R10, from once every 2 sec to once every 2½ min. Capacitors C2 and C3 charge to 20 V. When relay K1 activates, a negative charge is dumped into the battery.

Burping continues after the battery is fully charged. But the time between burps should be adjusted to correspond to the chemical activity in the battery—a short time for high activity when the battery charge is low, and long when near full charge. The duration of the negative

* Okolowicz, J., "Timing Circuit Burps Battery to Improve Charging Efficiency," *Electronic Design*, June 7, 1977, p. 116.

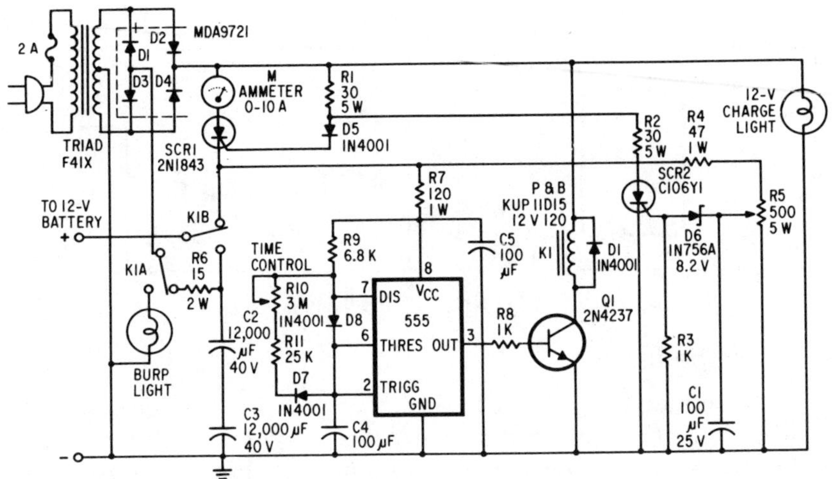

Fig. 8-10. Improved battery charger.

pulse is short, so it won't harm the battery. But be careful—with the battery mounted in an automobile, the alternator rectifiers can be damaged if the positive cable to the battery isn't removed.

8.10 neon lamp power supply

Very often it is necessary to light a neon in a circuit that would ordinarily not require any other high voltage. Neons are nice, because they can be used as indicator lamps and consume very little power. There are also a wide variety of lighted switches that come with neons built in. The supply of LED lighted switches is more limited, and more expensive.

In order to provide a small, inexpensive supply for the neon that can work off a small battery, the circuit in Fig. 8-11 was built. The circuit uses a 555 as an astable operating at a high frequency and uses a standard audio output transformer to step up the voltage. This offers no problems, because the current required to fire a neon is very low.

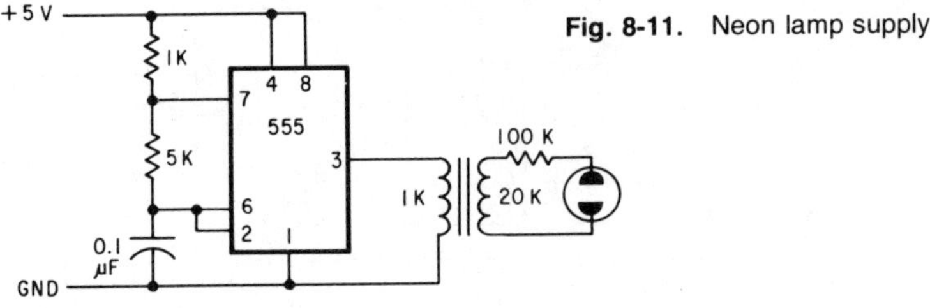

Fig. 8-11. Neon lamp supply.

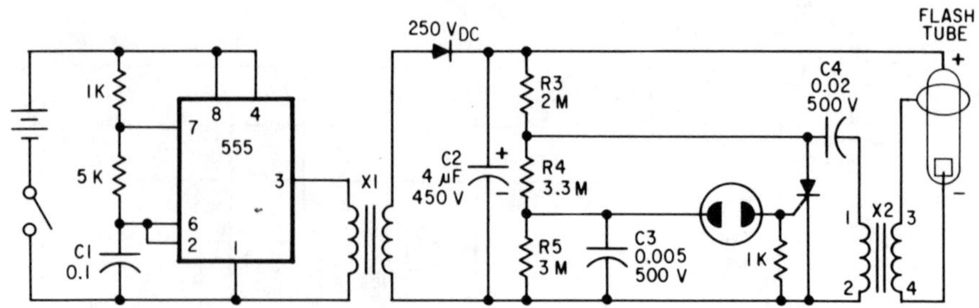

Fig. 8-12.   Strobe flasher.

8.11   strobe flasher

This circuit (Fig. 8-12) is similar to the previous one, except instead of using the high-voltage converter to light only a neon, it is used to fire a flash tube, much like those found in electronic flashguns for cameras. A bright flashing light of this type will make it a snap to locate your car at night in large parking lots. Put a red lens over it and it can be used as an emergency flasher to warn of a hazard.

Once the voltage is stepped up by the first transformer, it is rectified and charges capacitor C2 to about 250 V. The voltage divider composed of R3, R4, and R5 charges C3 to 90 V and C4 to 200 V. The voltages across C3 and C4 can be considered to be proportional to that across C2 because their time constants are small.

When the voltage across C3 reaches 90 V, the neon lamp fires, triggers the SCR, and causes the charge stored on C4 to be dumped into the primary of the trigger transformer X2. This voltage pulse causes a pulse of a few thousand volts to appear on the secondary of the trigger transformer, due to the high step-up ratio. The high voltage pulse in turn fires the flash tube, and then the cycle starts all over again.

8.12   power-failure alarm

If you take an astable multivibrator and force the timing capacitor to have the full charging voltage on it all the time, the astable will not

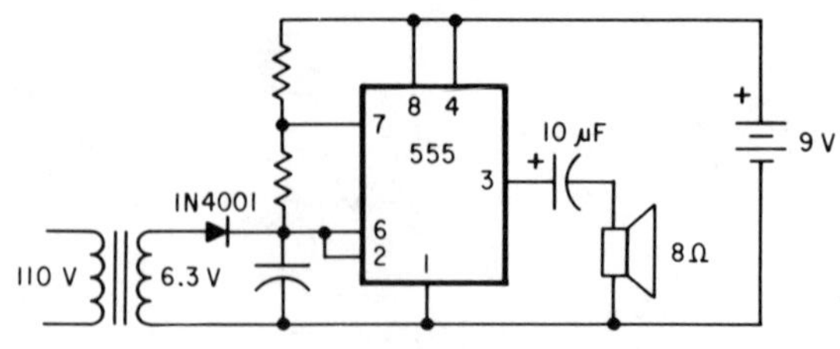

Fig. 8-13.   Power-failure alarm.

oscillate. It is this principle that is used in Fig. 8-13 to produce a power-failure alarm.

The astable is connected to a speaker on the output. A small 6.3-V filament transformer is connected via a diode to the timing capacitor of the astable.

The diode acts as a half-wave rectifier and supplies a charging current to the rechargeable battery as well as to the $0.01\text{-}\mu\text{F}$ capacitor, keeping them fully charged. When a power failure occurs, this constant source of voltage disappears, and the astable can operate normally and produce its warning tone.